A CRITICAL INDUSTRY UNDER ATTACK
THE STRUGGLE TO PRESERVE METALS AND MINERALS MINING VIABILITY IN THE U.S.

VOLUME 3

PINE NUT PRESS
Minden, Nevada

Copyright © 2018 Susan Lee Parkhurst

All rights reserved.

ISBN: 1978450028
ISBN-13: 978-1978450028

Library of Congress Control Number:
2017960235

The Dave W. Parkhurst Mining Writing Collection
was compiled, edited and designed
by Susan Lee Parkhurst
and produced by Pine Nut Press.

Note to Reader

The content of this book is the work of late mining writer/consultant Dave W. Parkhurst (deceased, 1993). The articles contained herein are reprinted from the *California Mining Journal* (now the *International California Mining Journal,* or *ICMJ)* with permission from the *ICMJ* publisher. No substantive changes have been made to the text in the process of digitizing the material for publication in this book, and no liability is assumed for any inaccuracies that may have occurred in the process of converting the printed magazine articles to the digital format, or from the digital format to the printed book. Further, the content reflects the state of the mining industry and its technologies and processes, as well as the knowledge base that existed, as of the time the articles were published in the *CMJ*. No guarantee is made as to the current accuracy or relevance of any of the content.

*To Skyler, Rody,
Cherie, Shamra, Mica,
and David*

Contents

Letter from the Governor vii
Editor's Preface ix
Editor's Acknowledgements x
Introduction 3

PART 1: ISSUES, CHALLENGES, AND THREATS
1   America Needs a Strong Mining Industry 7
2   Mining versus Environmentalism 11
3   Next Mining Boom, The 15
4   Placer Mining vs. Natural Erosion 21
5   Wilderness Issue, The 25
6   Environmental "Protectionism" 29
7   1987 Was a Good Year (?) 33
8   BLM Reclamation Bonding and the Small Miner 37
9   Miners Are a Threatened and Endangered Species 41
10 Wilderness Update: How Much Is Enough? 45
11 Mining Issues in 1993: What Can We Expect? 49
12 "FederalSpeak" Dictionary 61

PART II: MINERS VS. ANTI-MINING EXTREMISM
13 Extremists Mount Offensive in Environmental War 67
14 Environmental Activists Organize to Eliminate Mining Laws 73
15 Mineral Policy Center Supports Environmental Extremism 81
16 MPC Leads Effort to Abolish U.S. Mining Laws 87
17 Mineral Policy Center Launches Another Attack on Miners 91
18 Mother Nature Charged with Environmental Crimes *(satire)* 97
19 Anti-mining Propaganda: Half-truths and Lies 105
20 Miners Show Strength and Unity at Mining Law Hearing 111
21 Miners Dominate Reno Mining Law Hearing 115
22 Strong Opposition Shown Towards Senator Bumpers' S.433 119
23 Is "Activism" a Dirty Word for Miners? 121

PART III: DEFENDING THE 1872 MINING LAW
24 History of U.S. Mining Law, The 127
25 Sen. Bumpers' S. 1126 & Potential Impacts of Similar Legislation 131
26 H.R. 3866: Rahall's Mineral Exploration and Development Act 135
27 Rep. Rahall Introduces New Mining Law Bill 141
28 An Analysis of S. 433: Senator Bumpers' Mining Law Reform Bill 143

29	An Analysis of H.R. 918: Rep. Rahall's New Mining Law Bill	147
30	H.R. 1096 Would Restrict Use of the Nation's Public Lands	151
31	H.R. 2614: Rep. DeFazio Attacks the 1872 Mining Law	155
32	Rep. Les AuCoin Launches Assault on Miners	161
33	H.R. 918: Bill Would Stop Mining on Our Public Lands	167
34	Bumpers' "Amended" S. 433 Is a Trojan Horse	173
35	Recent Action on Mining Laws in Congress	179
36	Senator Bumpers Launches S. 257 Anti-mining Missile	183
37	H.R. 322: Rep. Rahall's "Mine Free By '93" Bill	189
38	Senator Craig's S. 775 Clears Senate Committee	195

Afterword 199
About the Articles' Author..200
Dave W. Parkhurst Mining Writing Collection (Appendix A) 201
Year of Publication in the *CMJ* (Appendix B)....................................203
Author's Comments on Bumpers' Bill S. 1126 (Appendix C)............205
Description and Analysis of Rahall's Bill H.R. 3866 (Appendix D)..211
References..215
Index...217
Editor Contact Information 223

Letter from the Governor

Editor's Note: As in Volume 2 of the Dave W. Parkhurst Mining Writing Collection, the content of this section replaces the Foreword for this volume. What follows is the text of a letter to Dave W. Parkhurst from Nevada Governor (Acting Governor, at the time) Bob Miller in August 1990 regarding Congressman Nick Rahall's proposed legislation, H.R. 3866, the "Mineral Exploration and Development Act of 1990." Dave was slated to provide testimony to a subcommittee in Washington, D.C., the following April in opposition to Rep. Rahall's proposal to "reform" the U.S. mining laws. (By then, a newer version of the proposal, H.R. 918, was before the Subcommittee on Mining and Natural Resources in the U.S. House of Representatives, and that was the bill Dave actually testified against on behalf of the Nevada Miners and Prospectors Association.) Governor Miller's letter to Dave encapsulates some of the significant legislative and regulatory issues that confronted the metals and minerals mining industry in the United States, especially in the American West, in the early 1990s. Such issues were chronicled in the articles contained in this volume of the collection.

Dear Mr. Parkhurst:

Thank you for your letter informing me of your plans to testify before Congressman Rahall's committee on the Mining Law.

Nevada knows the meaning and importance of a viable hardrock mining industry. There are well over 300,000 active mining claims in Nevada, representing approximately one-third of the hardrock mining claims in the nation. Nevada mines produce more gold and silver than any other state. Our industry also leads the nation in the production of several other mineral commodities. In addition to its importance to the economy of Nevada and other western states, domestic mining plays a critical role in reducing the need to import mineral products from foreign sources.

I am aware of the proposals in Congress to replace the Mining Law. Nevada's Commission on Mineral Resources and my office have gone on record as being opposed to those changes. We believe the proposals, including Congressman Rahall's H.R. 3866 [the "Mineral Exploration and Development Act of 1990"], increase the cost of holding and developing a claim substantially. The result will clearly be a reduction in mineral development activity—especially new exploration.

Nevada has made very significant progress in developing state law and regulations to ensure that mining is done in a manner which protects the environment and public safety. We believe these issues are best handled by state and existing federal regulations rather than through new federal statutory requirements.

For example, on August 28, the State Environmental Commission adopted regulations implementing the state mined land reclamation law which applies to all Nevada lands, regardless of ownership. The bonding requirement will assure the land is returned to beneficial use when mining or exploration is complete. The Bureau of Land Management has also instituted a bonding policy nationwide.

Nevada also has comprehensive mining permit requirements for protection of groundwater, air quality and wildlife. Again, these rules apply regardless of land status. Most western states regulate modern mining in similar fashion. Duplication at the federal level through placing land use and planning requirements in a mining claim location law is, in our opinion, not appropriate.

Good luck on your testimony and feel free to refer to this correspondence in your remarks.

> BOB MILLER
> (Former) Nevada Governor
> August 1990

Editor's Preface

This preface for Volume 3 of the Dave W. Parkhurst Mining Writing Collection adds to the background information provided in the prefaces to the other three volumes of the collection.

Dave W. Parkhurst was a respected mining writer, consultant, small miner and prospector, and fierce advocate for the industry he was proud to be part of. He had a lifelong interest in geology, minerals and mining stemming from his childhood, but his primary career until the late 1970s had been in the telecommunications field. In early 1980, when the price of gold hit $850 an ounce, everything changed. Long story short, Dave got into precious metals mining full-time that year and remained in the industry until he was felled by a massive heart attack one September day in 1993 while working in his home office.

Beginning in summer 1981 and ending with his death in 1993 Dave had been a writer for the *California Mining Journal*. During the 12 years of his association with the *CMJ*, hundreds of his articles had been published in the magazine. By the early 1990s, he had been named Associate Editor of the publication.

As Dave's widow I was devastated by his death and overwhelmed by my loss. In the years since he had died I had kept several boxes of the written materials in his files, being unwilling to simply discard what had been his life's work and represented so much of what he was about. The files included a copy of every issue of the *California Mining Journal* published since Dave first began writing. In late 2012, I received an email from a former administrator of the Nevada Division of Minerals inquiring whether I had any of Dave's written materials that I might want to donate to the division or state's archives. That got me thinking about producing an e-book of some of Dave's writing. Over the next five years the "e-book" evolved into a collection of nearly 150 of the articles that he had written for the *CMJ* compiled in a four-book, printed set.

Producing this compilation of Dave's work has been a far more massive project than I had envisioned at the outset, but it was something I was committed to doing and determined to complete. It was a way for me to honor Dave posthumously and to preserve his legacy. I hope he would have been pleased with this collection of his work.

Editor's Acknowledgements

My thanks to all those who helped in some way to make the Dave W. Parkhurst Mining Writing Collection possible. The specific individuals, whose contributions pertain to the entire collection rather than its individual volumes, are acknowledged in Volume 1, and the particulars of their contributions to this project can be found there.

A CRITICAL INDUSTRY UNDER ATTACK

THE STRUGGLE TO PRESERVE METALS AND MINERALS MINING VIABILITY IN THE U.S.

VOLUME 3

Introduction

As the title of this book indicates, the articles in the pages ahead primarily concern the vital role of the metals and minerals industry in the United States and its fight for survival. In these articles their author offers an impassioned and reasoned defense of this industry against the incessant, ferocious attacks it was under during the years he was writing about it (from 1981 until his death in 1993). He also issues a call to action by those most affected by the attacks and the attempts to severely constrict mining's ability to function in the U.S.: that is, miners and the mining community.

Volume 3 of the Dave W. Parkhurst Mining Writing Collection begins with a treatise on America's need for a strong minerals industry in Part I. This part covers a wide range of issues and entities that pose threats and challenges to not only the industry itself and those engaged in it, but to the country as a whole—our economy, our national security, and our way of life; in other words, civilization as we know it.

In Part II of Volume 3 the focus of the articles, with titles such as "Extremists Mount Offensive in Environmental War" and "Mineral Policy Center Launches another Attack on Miners," is on mining versus anti-mining extremism. One of the articles in this section is a satirical look at environmental regulations carried to extremes and a comparison between the effects of the natural forces on the earth and those created by human activity ("Mother Nature Charged with Environmental Crimes" on page 97).

Dave reported extensively on the political, legislative and regulatory actions of government, both federal and state, affecting mining. Part III of this volume begins with the history and development of the Mining Law of 1872, as amended. In the articles that follow this historical overview of U.S. mining law, the author chronicles the ubiquitous efforts to undermine or eliminate the law legislatively and the reasons why the mining law must be defended and preserved.

The reader is reminded that the events written about in these articles occurred well over two decades ago, so the content is obviously dated; however, much of it is relevant today.

A note here also about the way in which this volume of the collection is organized (specifically, the sequence in which the articles appear). In each of the three parts, the articles appear in chronological order by year of publication in the *California Mining Journal*. Appendix B in the back of the book is a table listing all of the articles in this volume and the corresponding year of publication. Being able to identify the year of original publication can provide some historical and political context for the events chronicled in this volume and Volume 4, the final volume in the Dave W. Parkhurst Mining Writing Collection (*Fighting the Good Fight: Mining's Battle for Survival in the American West*).

PART I

Issues, Challenges, and Threats

"As a rather explicit analogy, the U.S. mining industry is now living in the equivalent to the Paleocene Epoch (early Tertiary), during which the earth saw the extinction of dinosaurs and ammonites and the development of flowering plants. All we have to do now is step into the tar pits so our current development is preserved intact for examination in the future."

1
America Needs a Strong Minerals Industry

MOST PEOPLE IN THE UNITED STATES do not fully realize just how important mining is to maintaining the basic foundation of modern civilization and providing the materials necessary for each and every person's daily requirements. It is unfortunate that Americans have become so accustomed to having an abundance of material goods available that the knowledge of their importance in our daily lives has been largely ignored or forgotten. The nation's basic reliance upon metal and mineral products has grown to such an extent that modern civilized society would collapse if these materials were no longer available.

Government figures show that the United States consumes about one-fourth of all the minerals produced worldwide, yet the country only has about 5 percent of the world's population and only covers roughly 7 percent of the planet's total land area. Recent statistics indicate that we use about 40,000 pounds of new mineral products for every man, woman and child in this country alone every year.

In addition to agriculture, mining is the other major basic industry that provides a basis for the creation of our country's wealth. All of our other domestic industries are ultimately dependent upon the mineral products produced by mining. Even our food supplies are dependent upon mineral products to fertilize, plant, harvest and distribute crops and on the equipment needed to process them into a marketable form.

The raw materials provided by mining are absolutely essential to our civilization, yet some people persist in creating conditions that weaken the industry that produces them, as well as the industries that utilize the mineral materials to produce finished goods for our basic needs. A number of factors are involved in the gradual weakening of the nation's mining and manufacturing sectors, but the most important factors are the active opposition of radical special interest groups and a singular lack of public awareness or interest.

Over the past 20 years or so, many Americans have become preoccupied with self-interests and environmental concerns to the point where they have lost touch with the nation's basic priorities. Little or no attention is being given to the underlying importance of U.S. industry to our economic stability and personal well-being. The major thrust of public opinion has instead been directed towards activities that threaten to undermine our basic security, standard of living and individual freedom.

This nationwide change in attitude has resulted in a constant clamor for overregulation of business and industry, excessive and costly litigation, demands to close off huge areas of the public lands to natural resource development, and an unrealistic, and generally hostile, attitude towards the business sector—all of which have served to further weaken our mining and manufacturing industries. This situation poses an interesting question: What would life be like if no minerals were produced?

Without mining, there would be no transportation available— unless we were to return to using horses, ox carts and sailing ships. There would be no cars, trucks, buses, railroads or aircraft. Each of these modern forms of transportation is totally dependent upon metals, metal alloys and mineral products for their basic construction. A partial list of the metals utilized in the manufacture of automobiles includes iron, tungsten, molybdenum, aluminum, copper, zinc, lead, tin, antimony, manganese, cobalt, nickel, chromium, tellurium, titanium and vanadium. Solid state ignition systems and catalytic converters also use such elements as germanium, arsenic, silicon, silver and the platinum group metals. Many other metals, alloys and mineral compounds are utilized for specific component parts, paints, ornaments and safety devices.

Without mining, there would be no electrical power. Electrical power generation and distribution is also totally dependent upon metals, alloys and mineral compounds. There would be no electricity to operate machines and appliances in factories and homes. A partial list of metals used in electrical power generation and distribution includes copper, aluminum, iron, cobalt, nickel, lead, zinc, tin, gold, silver, beryllium, cadmium, germanium, indium, manganese, mercury, molybdenum, tantalum, titanium, tungsten, the platinum group metals, and others. All of these metals were originally discovered, extracted and refined by the mining industry.

Electrical power and transportation provide good examples of some basic needs for metals and minerals, but what about modern appliances, machinery, tools and, literally, the nuts and bolts that hold almost everything together? No available metals means: no refrigerators, washers, dryers, stoves, furnaces, kitchen appliances, light bulbs, television sets, stereos, tools, computers, typewriters—the list goes on and on. It also means that the machinery necessary to produce these items could not be built. As a consequence, there would be no soaps, shampoos, cosmetics, razors, textiles, manufactured clothes, packaged or canned or bottled foods, or, for that matter, toilet paper.

Without the mining and manufacturing industries, everything would have to be made by hand and no effective system for the distribution of goods could possibly function. So, at a time when we are totally dependent upon the finished products and goods provided by the industry, why is our country becoming less self-sufficient in the production of mineral raw materials?

The United States has legislated and regulated itself into a position where we can no longer compete effectively with foreign producers. And what has been the result of these actions? The U.S. has exported several million high-paying jobs to foreign countries; our mining, processing, smelting and refining capacity has been sharply reduced; foreign suppliers now dominate several commodity markets in which the U.S. was once self-reliant; many major U.S. industries are relocating overseas; our base metal mining and processing industries have undergone a severe decline; the U.S. is piling up foreign debts at an astronomical rate, and the nation has rapidly become a net supplier of "services."

At the present time, the United States is totally dependent upon foreign sources for many of the metals previously mentioned and it is reliant on foreign imports for the majority of its supplies of many other metals and minerals. We are also dependent upon foreign suppliers for most of our basic fuel and oil requirements. Our nation is becoming more and more dependent upon foreign sources of supply for its raw materials almost daily while, at the same time, our ability to produce these basic mineral commodities is still declining.

What would happen to us if these sources of supply were suddenly restricted or cut off entirely? Simply put, our country could not sustain itself for any appreciable length of time.

Our national security and economic viability are rapidly becoming contingent upon the good will and reliability of our foreign trading partners. In many instances, we obtain the major portion of several critical and strategic mineral commodities from foreign nations that are either politically unstable or are basically opposed to our beliefs and ideals. What will occur when the U.S. becomes totally dependent upon these sources of supply?

At best, we could expect another OPEC-style confrontation that would drive commodity prices through the ceiling—just as occurred during the energy crisis of the mid-1970s. At worst, our supplies could be cut off entirely and we would face critical shortages.

Contrary to popular opinion, this nation's industries could not turn about immediately to begin producing mineral commodities and finished goods if they were no longer available from foreign sources, or if the foreign suppliers suddenly refused to advance additional credit. The lead time required to discover, evaluate and develop a mineral deposit or to redirect manufacturing capacity is usually measured in terms of years. In today's world, we would begin to suffer critical shortages of necessary goods and materials within months.

We need to turn this situation around by implementing policies that actively encourage our domestic mining and processing industries, rather than discouraging them. Mining and the production of finished products are critically important to our very survival as a nation, as well as being necessary to maintaining our way of life.

America needs a strong minerals industry.

2
Mining vs. Environmentalism

IT SEEMS THAT ANOTHER ALL-OUT EFFORT is being made to close off additional massive tracts of public lands to any type of natural resource development. Having been notably successful in their previous efforts to lock up this nation's federal lands in a private preserve for a small segment of the population, environmentalists are now engaged in the final struggle to get it all. In the process, they may permanently disable our capacity to become self-sustaining in the production of minerals and other natural resources.

Though difficult to imagine, this final push to make the U.S. into a massive national wilderness system is occurring at a time when our industrial strength is approaching an all-time low. The nation is currently in the process of setting record foreign trade deficits, with import levels of foreign commodities reaching a corresponding all-time high. The end result of this incredible public fascination with "environmentalism" is the ultimate destruction of this country's ability to defend itself during any prolonged national crisis. Even if such an emergency does not occur, this type of activity is making the U.S. increasingly vulnerable to international blackmail by those foreign countries supplying our critical and strategic mineral and energy requirements.

The foregoing statements might seem rather strong, but consider what has happened historically to any nation which has been deprived of its supplies of raw materials and/or energy-producing fuels. In every case, these countries have either been transformed into second- or third-rate powers—or they have been conquered by other nations. With the ever-changing course of international politics, there are, in reality, no longer any permanently "safe" sources of supply for our mineral and energy requirements—with the exception of Canada, and possibly Mexico. All other foreign sources of supply could be disrupted very easily by an unfriendly power, if not disrupted by an internal political shift within the country or countries supplying certain critical mineral or energy commodities.

Many well-intentioned conservationists will disagree with the critical nature of this assessment, but they are, regrettably, not at all familiar with the nuts and bolts of mineral and energy development and production. On the average, most environmentalists also have a totally unrealistic view of the international political scene and its potential effect upon the basic survival of the United States. Far too many honest individuals have become enamored of the all-or-nothing philosophy being preached by a few radical members of their respective groups. Whatever happened to the concepts of multiple use and reasonable compromise?

The leaders of this nation have already realized that environmental degradation and pollution must be contained and corrected, and they have passed a multitude of laws, rules and regulations to accomplish this objective. At no time in our past history have we been more aware of these problems and the necessity of their rapid solution. Many comprehensive programs have already been initiated, and a sizable portion of the federal budget is being devoted to these purposes. As with the culmination of any worthwhile effort, however, we are again going for overkill.

Our society has become so accustomed to an overabundance of material goods that it has been intellectually divorced from the reality of exactly where those goods come from. And no consideration is being given to the fact that our nation requires a healthy business and industrial environment as well as a healthy natural environment. The more affluent a society becomes, the greater its tendency is to overlook those things that are necessary to its basic survival. The pioneer attitude of "How are we going to make or provide these things?" has been replaced with the idea that we can always get what we need "somewhere." The problem is that "somewhere" is not always a viable option, and it often does not even exist.

The effective and responsible management of our natural resources involves not only the protection of those resources, but also the careful and balanced use of same. It does little good to lock up those resources that are vital to our national security in the interest of "protecting" the surrounding environment. If our country does not survive, just whom are we protecting the environment for?

Most of the people in this country hold the belief that we can always make any necessary changes in policy to allow for unforeseen circumstances, including the immediate use of our natural resources in the event of an emergency. This type of thinking can lead to a fatal mistake, because of one commonly overlooked factor: *time.* If a national emergency were to occur under the conditions currently existing in the world today, the most time we might have available to us would be measured in months. The effective production of natural resources, particularly minerals, requires lead times that are measured in *years*—even if the source of material is known and defined, let alone the time required to discover new sources of minerals in areas that have been withdrawn from mineral exploration.

Another tragic consequence of massive wilderness withdrawals is that they are most often made in areas that are the most favorable for the deposition of minerals. The same geological conditions that create the scenic mountains which most environmentalists seek to protect also create ideal conditions for the formation of mineral deposits. As a consequence, the areas representing the highest potential for natural resource development are the first to be closed off.

The general public's awareness of the basic issues involved in wilderness withdrawals is, as usual, extremely limited and biased. Because of the enormous amount of publicity that environmentalism has been given over the past few years, very few people are really aware of what is actually going on. The news media has been, and still is, very reluctant to publicize the other factors involved for two basic reasons: an uninterested public and the high popularity of the environmental movement. As a result, only those publications directly concerned with the critical nature of the problem publish any worthwhile information on the subject—for an unfortunately limited readership.

One of the main reasons why the more radical members of the environmental movement are now pushing the wilderness issue to such extremes is that they also realize their distinct advantage due to the "information gap" between proponents of both sides of the issue. So they feel that now, while the public is still basically misinformed, is the best opportunity they have to close off the maximum amount of public lands to development.

And, who stands to gain the most from tying up tremendous acreages of our nation's most productive natural resource areas? It is not, unfortunately, our own future generations of Americans, but rather a few self-serving individuals and this nation's enemies. A number of foreign governments would like nothing better than to permanently undermine the industrial and economic strength of the United States.

We must bring this mania for ever-increasing amounts of wilderness lands to a screeching halt, before it is too late. As mentioned above, we no longer can afford the luxury of sitting back while this country becomes one large wilderness preserve. "Environmentalism" has gone unhindered for too long, and it is now threatening the security of our country.

Author's Note: This article was written with the express purpose of calling attention to the totally unreasonable and unrealistic demands being made by the more radical elements within the environmentalist movement, and it is not meant to detract in any way from the legitimate concerns for the adequate protection of our environment. I am an outdoorsman and conservationist myself, as well as a miner.

3
The Next Mining Boom

THE MINING INDUSTRY TYPICALLY undergoes periodic boom-and-bust cycles due to a recurrent imbalance between supply and demand, and the present cycle has obviously been going in a downward trend for the past few years. This downward trend largely results from the recent economic recession and consequent reduction in demand for mineral commodities—which, in turn, depressed the market prices for most mining products. Under normal circumstances, an increase in industrial output and demand would have worked its way down to the mineral industries and produced the next upswing in mining activity. This time, however, several new factors have influenced and delayed the recovery of the mineral-producing sector.

Worldwide financial and economic policies over the last decade have been directed towards the development of basic industries in the Third World countries—including the production of mineral commodities. As a result of these policies, many new foreign mines have come on line at the same time that the world experienced an economic recession. This lack of foresight produced a new phenomenon. At a time when demand for minerals was sharply reduced, with a consequent overabundance of minerals available on the markets, worldwide production of these minerals actually continued to *increase.* This had a notable effect upon an already depressed market, as it depressed metal prices much further than under normal supply-demand conditions and, since the oversupply continues to hang over the market, most of the mineral prices are remaining below the cost of production. As a result, many of the larger base-metal mines in the U.S. and Canada were forced to make drastic cutbacks or shut down indefinitely.

Then came an even bigger shock: most of the major mineral-consuming industries in North America did not recover as the economy improved. This was also due to the development of Third World industry, as many new plants that produce finished metal and mineral products were also coming on line. There are now a large number of cheaper, finished foreign products on the world markets, most of which are being sold at prices below the cost of production by industries in North America. This means that many of the basic industries in America remain depressed, and the demand for minerals has not increased sufficiently to overcome the overabundance in supply. It also means that a large number of jobs and a good part of our economic strength has been literally exported to foreign competitors. Unfortunately, these conditions are very likely to remain as a depressant on the mineral sector, because most of the foreign production is subsidized by the various governments concerned and these nations cannot afford to cut back on production—largely because of the tremendous debts incurred in the development of these same basic industries.

As a result of the abovementioned conditions, the mining industry in North America has reached a critical turning point in its basic direction. Many of the major mining companies have already started a switch to the precious metals, and more are rapidly starting to diversify their mineral holdings. A large number of big mining projects have been placed on indefinite hold, particularly those involving certain base metals normally in demand in the steel industries. Drastic cutbacks have been made in the production of copper, tungsten, molybdenum, barite and other minerals. This trend is expected to continue.

So, what direction is the mining industry in North America going to take now, and in the future? According to our economic and political leaders, a new high-tech boom is on the horizon—as a major switch is made from the past basic materials-producing industries to a new emphasis upon the production of finished products utilizing space-age technology. In other words, it is predicted that our industrial base will swing from the production of steel and construction materials to the manufacture of computers, satellites and other sophisticated electronic devices. Such a shift in our basic industries would, of course, have a major impact upon the demand for the minerals and mineral products produced by the mining industry, and also upon the *type* of metals and minerals produced.

The current problem is to accurately define which metals and minerals will be needed by industry in the future and also to approximate the relative quantities required. As the probable future direction of the manufacturing industries is not yet ascertainable, it is very difficult to estimate their basic material requirements. As a result, the processes of future mine planning and development have largely been put on hold because of a lack of sufficient information. One thing is certain, however: the new structure of the mining industry will be considerably different than it has been in the past.

This state of flux, or change, in the industry is one of the major reasons behind the pronounced shift towards precious metals production. The emphasis being placed upon precious metals exploration also says a lot about the future potential for these metals. Most of the larger mining companies would not diversify into gold and silver properties unless they were fairly certain that future prices and demand would justify the move. A good indication of the future stability in demand for both gold and silver can be seen in recent statistics (1983) on their end-use consumption. For gold, 55% of total consumption was for jewelry and 33% in the manufacture of electronic devices (88% of all U.S. end use). For silver, photography used 43%, electronics 26%, and jewelry and electroplating 15% (84% of all U.S. end use). These uses can be expected to continue and gradually increase, and, as the basic industry shifts into high-technology production, the electrical and electronic end use for both metals should undergo a dramatic increase. This indicates that a substantial part of the next mining boom will involve precious metals exploration and development.

Numerous other metals are utilized in high-tech type hardware and software, though not as yet in appreciably large quantities. Any significant increase in the production of space-age electronic products would, however, certainly result in increased demand for two specific groups of metals: the platinum metals and the rare earth metals. The platinum group includes platinum, palladium, rhodium, ruthenium, iridium and osmium; the rare earth metals include lanthanum, cerium, praseodymium, neodymium, promethium, samarium, europium, gadolinium, terbium, dysprosium, holmium, erbium, thulium, ytterbium, lutetium and, usually, yttrium. End-use consumption of the platinum metals in 1983 for automotive, electrical, chemical and dental uses was 90%, and any increase in production of high-tech devices would produce a greater demand for these metals. The same type of increase in demand for the rare earths would also probably occur, as 85% of end-use consumption is currently for the petroleum catalyst and metallurgical applications.

As the elements within each of these separate groups commonly occur together or in association with one another, it is expected that a significant portion of future mining exploration and development will be directed towards the production of these metals and several of their compounds. It is also expected that further research into possible new applications for these elements could result in an additional increase in demand.

Leaving aside the potential effect of high-tech demand for the moment, it is also logical to assume that this country will make an attempt to become more self-sufficient in the production of several metal and mineral commodities upon which we are now import-reliant. Most notable among these mineral commodities are bauxite and alumina, cesium, chromium, cobalt, columbium, corundum and diamonds, manganese, mica, nickel, strontium, tantalum, thallium, thorium, tin and yttrium. Since the exploration for, and development of, these minerals is practically essential to our national security, a significant shift in priorities related to minerals policy by our government is expected to occur in the near future. As can be seen by the number of commodities listed, this might produce a considerable shift in the priorities of our mining industry.

Since most of the known ore deposits containing these minerals are fairly low-grade, at least in the U.S., it might be necessary to provide price supports or other incentives to encourage their development. In addition, adequate smelting and refining facilities are sadly lacking in the United States, and it might become necessary to establish more realistic environmental standards that would allow these industries to operate. At present, this country exports much of its ores and concentrates to foreign countries for smelting and refining—largely due to unrealistic environmental standards and an uninformed public. If this condition is allowed to remain unchanged, it would be a practical impossibility for the U.S. to even remotely approach self-sufficiency in the production of these minerals.

Even in the face of certain attempts to obstruct progress in this area (and since necessity is the mother of invention—and compromise), it is reasonable to expect that added emphasis will be placed upon the domestic production of these minerals. Therefore, it is expected that a fairly large portion of the future mining activity in this country will be devoted to development of domestic sources of supply for those mineral commodities that we are not currently producing or processing.

It is also necessary to implement advanced technology in the mining and processing of mineral materials in the U.S., in order that the mining industry might withstand foreign competition. This could involve a considerable amount of retooling and increased automation in extraction and processing methods, as well as cost-cutting moves in other areas of operations. This type of change will tend to produce more streamlined operations and could, in turn, create a minor high-tech boom in the mining industry.

In summary, it appears that the most notable feature of the next mining boom will lie in its diversity. It seems likely that a greater emphasis will be placed upon the recovery of larger quantities of those metals and minerals required for an advanced technology, with a consequent reduction in demand for most base metals used in construction materials. It also seems evident that precious metals mining has an assured future.

4
Placer Mining vs. Natural Erosion

THE PUBLIC'S PREOCCUPATION WITH environmental concerns has again resulted in additional unrealistic restrictions on placer mining activity. Foremost among the most extreme examples of overregulation are the Environmental Protection Agency's (EPA) proposed standards for the maximum allowable discharge of particulate matter from placer operations into streams and rivers.

Under the present water quality standards considered acceptable by the EPA (and also by many state agencies), Mother Nature could be subject to heavy fines (and a court order closing down her operations) for polluting most rivers and streams, over a good part of each year. It is doubtful, however, that Nature will allow herself to be regulated so harshly. Insofar as placer miners are concerned, there seems to be no available recourse to alleviate this new burden.

It is interesting to note that the particulate standards were purportedly devised to eliminate the deleterious effect of such "pollutants" on fish, wildlife and water quality. Several studies are reported to have shown that higher levels of particulate matter have an adverse impact upon fish and wildlife. How, then, do the fish and wildlife survive Nature's onslaught—on a magnitude that defies any comparison to disturbances resulting from placer activities? Keep in mind that we are referring to *natural* substances as pollutants, and that these materials have been present in the environment since life began on this planet.

For purposes of comparison, consider that several other studies have shown that natural erosion in our river and stream systems moves much more such particulate matter *each year* than has been displaced by placer mining activity since it first began. One major thunderstorm can move more material in a watershed over a few hours or days than could be released by placer mining in a year. Each year's spring runoff in the same area would transport such a large volume of material that placer disturbances would not even be of significance by comparison.

Why, then, is there so much attention being given to disturbances caused by placer mining? Some of the reasons for this are: the high visibility of placer operations (they attract attention); an extremely localized viewpoint on the potential effects of pollution; no rational comparison to natural erosion, and problems that occurred in the past but no longer exist today ("Just look at what happened with hydraulic mining!"). Another major influence lies in self-righteousness ("I'm right, you're wrong, and I won't listen to you!"). In the absence of reason, there is no possibility for compromise.

Placer operations are highly visible, and it is therefore very easy to find fault with them. If you are looking for ways to find something wrong with a particular situation, you will. Many people live with extreme pollution in their immediate environment but, in the absence of a sharp contrast, they hardly notice it. In a clean and natural environment, however, anything unnatural can be called pollution.

This is where the "localized" viewpoint comes into play. The amount of particulate matter being discharged into a stream is usually measured at points a certain number of *yards* downstream from a placer operation, and compared to the water purity immediately above the site. In most cases, very little of the material has had a chance to disperse or settle out. The *average* amount of particulate matter present in the entire watershed throughout each year is not even considered. Nor is the amount of suspended materials present during periods of high runoff taken into account.

There is, admittedly, a need for some type of control on mining activity to ensure against the degradation of the environment. We only have to look at the past to see what can happen when there are no regulations in effect. But, does that mean that we have to go to such extremes in the attempt to prevent abuses?

A much more logical approach would be to monitor the average amount of particulate matter being discharged each year in a particular watershed, at a point well removed from any unnatural disturbance. If it was found that upstream placer activity produced a significant increase in the average, then controls could be instituted to remedy the situation.

Under the existing testing requirements, even a relatively small placer operation could be shut down for failing to meet stringent water quality standards. However, if the particulate discharge were measured in a more realistic and unbiased manner, it could be shown to have an insignificant effect when compared with the average natural particulates present in the watershed when no mining activity was taking place.

A few environmentalists maintain that the disturbance of material by placer activity is an "unnatural" action, and that the material would normally remain in place if not subjected to movement by mining. This concept is ridiculous, as the material would not be there in the first place if it had not already been moved previously (many times) by natural forces.

Stream placer deposits are formed by the relatively constant movement and sorting of gravels over extremely long periods of time. Without this continual movement and gradual concentration of valuable minerals, economic placer deposits could not form and there would be no reason to mine them. That this process involves the movement of enormous amounts of gravels is implicit in the fact that the material has been smoothed and polished by constant abrasion — even at great depths. If the movement of small quantities of gravel by mining is to be considered a source of pollutants, then the process of natural erosion must be regarded as a massive degradation of the environment. In this light, most of our scenic rivers and streams are naturally polluted to an incredible extent.

Thus far, we have attempted to establish that man's ongoing quest for placer gold, and other minerals, cannot possibly disturb the environment to an extent comparable to that accomplished by natural erosion. For that matter, it has been estimated that the total surface disturbance caused by mining (of all types) throughout the world amounts to less than two-tenths of one percent of the planet's land surface. It would be interesting to compare this with the amount of surface disturbance caused by construction of roadways or, for that matter, cities and towns.

There are actually some aspects of placer mining, particularly the use of dredges, that can serve to enhance the habitat of fish and wildlife. For instance, the "pot-holing" effect of placer dredges provides small pools in a streambed that can be used by fish and wildlife — especially in intermittent streams or during dry years. In some areas, fish can survive in these pools where they would normally die as the stream dried up. Fish have also been found to collect immediately below placer mining operations to feed on water insects dislodged by the movement of the gravel. If the mining activity had an adverse impact upon the fish habitat, they would go elsewhere.

Also, in drier climates and during dry periods in wet climates, stream gravels tend to become compacted or "cemented." Placer activity loosens the streambed and provides good locations for fish to spawn.

Another interesting fact is that wildlife tend to stay near mining operations, both to feed and to utilize the additional water available. They are seldom disturbed (or endangered) by the mining operations and would obviously leave the area if they were bothered. A large number of miners also care about fish and wildlife and exercise careful mining practices to ensure that the habitat remains undamaged. Many operations exert special efforts to *enhance* the wildlife habitat.

Our geological and topographical environment is not fixed in place; it is being constantly altered and changed by natural forces. It is now very difficult to detect signs of disturbances in areas that were mined extensively as little as 50 years ago. This is particularly true in the case of river and stream beds, as natural erosion tends to modify them considerably in a very short period of time. The use of terms such as "environmental degradation" in reference to streambed placer mining is ridiculous. Mother Nature will move the same material if we don't—and on a much more massive scale.

5
The Wilderness Issue

THE CONSERVATION AND PRESERVATION of natural or sensitive areas in our environment is both desirable and necessary. It is also absolutely essential that we drastically reduce the harmful pollutants that our modern civilization produces and subsequently discharges into the environment. It is not necessary, however, to halt the development of our natural resources and slow the progress of civilized society in order to achieve these objectives.

The entire wilderness issue arose as a result of widespread, and justifiable, concerns that the ecology of the world was endangered and that the ultimate survival of life on this planet was, and still is, at stake. As a result, public awareness and concerns have brought about a tremendous effort to address these problems and to provide corrective action. A multitude of laws, rules and regulations have been devised and enacted by federal, state and local governments with the dual objectives of correcting past abuses and preventing further degradation of the environment. The implementation of these policies is proceeding at an accelerated pace, and impressive results have already been obtained. Why is it, then, that we cannot continue to pursue our objectives in a reasonable manner, but are instead now resorting to unrealistic extremes?

Public attention is currently being focused, by both the news media and environmental groups, on the issue of additional "wilderness" withdrawals as a purported means of "protecting" the environment. According to the most vocal of these wilderness advocates, our very lives are at stake in the battle against "development interests." But is this really true? Or are certain special interests seeking to utilize public sentiment to gain their own ends?

To illustrate how ridiculous the "critical" nature of the wilderness issue actually is, let us look at several interesting statistics. In May 1982, the U.S. Senate Committee on Commerce, Science and Transportation published a report that, on page 140 of "An Assessment of Factors Affecting Small Mining & Custom Milling & Smelting Operations in the Western U.S.," reads as follows:

The availability of land for mineral exploration and development is an item of critical importance to the small miner. The initial intent of the 1872 Mining Law was to provide the resources of the land for unrestricted use in developing locatable mineral commodities. Over the years, the concept of unrestricted use of land for mineral development has undergone radical change due to the growth of population and the concomitant increase in awareness of environmental values. These values are clearly portrayed in the multitude of environmental laws and regulations which have been enacted by Congress.

The environmental movement has had a significant impact on the small miner in terms of the environmental regulations which have been imposed on mining operations. Furthermore, various Congressional acts have withdrawn large public land areas from mineral exploration and development. The result is that the small mine operator may now explore and conduct mineral development activities only on public lands not subject to these withdrawals.

In order to evaluate the level of restrictions inhibiting mineral development on the nation's public lands, the Secretary of the Interior established the Task Force on Availability of Federally Owned Mineral Lands in 1976. This task force reported that nearly 60 percent of the nation's public lands were formally withdrawn or highly restricted to mineral exploration and development under the mineral location and leasing laws. According to recent statements issued by Department of the Interior officials, approximately 70 percent of the nation's public lands are now closed to mineral development.

It should be kept in mind that this report was published in 1982, and that many more areas of the public lands have been withdrawn since that time. When considering these facts concerning the actual amount of public lands *already closed* to development, does it seem realistic, or desirable, to demand the additional withdrawal of *even more* areas as "wilderness" preserves?

The question now arises: Just how much is enough? The answer provided by most environmental groups is evidently *100 percent*. At a time when this country is faced with critical shortages of raw materials and is becoming increasingly dependent upon foreign imports to provide enough material to meet its basic needs, attempts are being made to achieve objectives that will further compound the problems by removing access to even more of our natural resources. In this sense, the demand for more wilderness is a direct threat to our national security.

Whatever happened to reasonable compromise? *Well over 70 per-*

cent of our public lands are already closed to resource development.

Surprisingly enough, the "multiple use" concept and "balanced" natural resource management programs presently being pursued by our two primary federal land agencies, the U.S. Forest Service and the Bureau of Land Management, are designed to achieve practically the same objectives as those propounded by the most vocal wilderness advocates. Both agencies are actively seeking the elimination of undue environmental degradation and are also promoting programs that will *enhance* the environment. It is becoming increasingly evident that the most effective means of preserving and protecting our public lands is through proper land management policies. *And these policies are already in effect.*

Where, then, lies the "critical" nature of the wilderness issue? It should be obvious that adequate steps are being taken to protect our public lands, so the wilderness issue must be designed to accomplish something else. Or, more likely, present demands for more wilderness are a result of our tendency to go to extremes and/or the lack of an informed public.

There are also several interesting statistics concerning those areas that have already been designated as wilderness tracts. It has been shown that once an area is designated as wilderness, it then attracts more public attention than it did previously and, as a consequence, more visitors. This, in turn, results in an added accumulation of discarded trash, the chopping-down of trees, and a notable increase in the incidence of forest and range fires. The fire factor is *very* important, because large areas of pristine forests have been burned off. Some of the worst fires occur in wilderness areas because, due to the absence of roads or other access, fire fighters are greatly hampered in their efforts to control even the smallest blazes. And many of these fires are caused by man.

It should be noted here that, in addition to wilderness areas, we also have a multitude of National Wildlife Refuges, National Parks, National Recreation Areas, Wild and Scenic Rivers, and other public land withdrawals. We are, in fact, the most conservation-minded society on the planet. Let's face it: we really don't *need* more wilderness areas, we only need to properly manage what we already have.

6
Environmental "Protectionism"

HERE WE GO AGAIN! ELATED and encouraged by their previous success, the environmental extremists are currently engaged in another campaign to lock up even more of our nation's natural resources. Bearing the banner of "environmental protection," these radicals are spreading misinformation, disinformation, half-truths and outright lies in an attempt to discredit and diminish the natural resource industries in the United States—most particularly, the mining and minerals-related industries.

In pursuit of their quest to make this country into one huge, nonproductive national park, these self-appointed saviors of mankind have left reasonable compromise behind. Rallying many confused politicians to their cause, they are successfully promoting legislation to withdraw massive amounts of the public domain lands for their personal playground. And, having achieved their initial objectives by undermining the public's confidence in our so-called smokestack industries, they are now even abandoning their previous supporters as they seek to prevent the use of our public lands for any productive purpose.

The people who now find themselves on the protectionists' hit list include miners, loggers, oil producers, energy developers, manufacturers, recreationists, farmers, ranchers, hunters, rock hounds, motorcyclists, four-wheelers, senior citizens and the handicapped. *No one* is to be allowed to interfere with the pristine solitude of the nation's great outdoors—excepting, of course, the environmentalists themselves.

And the radicals' objectives are by no means confined to the designation and permitted use of "wilderness" lands and other administrative land withdrawals; they are also attacking any potential development activity on nonsensitive lands, both public and private. Costly and time-consuming Environmental Impact Studies are the order of the day, often preventing or precluding even the simplest development projects by reducing their economic feasibility. Stringent environmental standards, special use permits, licensing requirements, civil litigation, bonding costs, political pressure and public opinion are often used as a means of preventing or discouraging new development activity rather than as a means for regulating it.

Organized, articulate and well-funded environmental groups are also attempting to control our nation's land management agencies by pursuing litigation in civil and administrative courts. Dissatisfied with the country's multiple-use concept for managing our natural resources and the use of public domain lands, these powerful groups are successfully preventing the Bureau of Land Management and the U.S. Forest Service from properly implementing the land management policies approved by Congress. This new program of perpetual litigation effectively ties the hands of duly appointed land managers, and effectively places millions of acres of public lands in de facto withdrawal status. As such, these lands will remain in a "protected" category until the environmental groups can find the means whereby they can be placed in a permanent withdrawal classification.

In a total disregard for rationality and common sense, environmental activists are presently asking Congress to:
- Abolish the 1872 Mining Law and substitute a mineral leasing system.
- Halt the construction of access roads into isolated areas.
- Reduce timber sales and harvests by one-half.
- Reserve water rights in or adjacent to designated wilderness.
- Create "buffer zones" around most wilderness areas and wildlife refuges.
- Permanently withdraw our offshore coastal areas from petroleum exploration and development.
- Withdraw huge additional amounts of the nation's public lands for wilderness designation.

Does this leave any doubt as to their ultimate objective?

By a calculated expansion of "environmental concerns" into another area, many property owners are finding themselves in the ridiculous position of being unable to build upon, or in any way develop, their own private property because of objections expressed by environmental groups. It has come to the point where anyone and everyone is allowed to voice their objections to any proposed development, regardless of whether or not they are, in fact, affected in any way by the development in question. In this type of scenario, a few vocal radicals may effectively block a proposed project—or make the process so expensive and time-consuming that the idea is eventually abandoned.

Where does this type of extremism stop? And what will happen if it is allowed to continue? And finally, who will stand to gain the most if our nation's natural resources are placed off-limits and progressive development in this country is brought to a halt?

Obviously, this country's enemies would like to see us continue to be deluded by an environmental mania that disrupts both our national security and economic strength. They would also rejoice if we continue to proceed with the overregulation and eventual destruction of our natural resource and basic industries.

The environmental protectionism mania that is currently sweeping the country is contributing significantly to our increasing reliance upon foreign sources for raw materials and finished goods. As a result it is weakening our economic stability, undermining our national security, and making us even more vulnerable to disruptions in supplies of critical and strategic materials.

If we don't stop this insanity before it goes any further, our nation will eventually find itself in a perilous position—particularly in the event of a national emergency.

7
1987 Was a Very Good Year (?)

MOST YEAR-END SUMMARIES usually provide a kind of wrap-up of the events, issues and trends that occurred, arose or became evident during the preceding 12 months. An attempt is normally made to accentuate the positive, downplay the negative, and generally try to view the overall picture through rose-colored glasses. This time, however, I believe that another approach might be more appropriate, since a healthy dose of reality can be beneficial from time to time.

For the minerals and other natural resource industries, 1987 produced a mixed bag of the good, the bad, and the ugly—unfortunately, mostly the latter. On average, the trend towards overregulation of mining and miners continued with a vengeance, the preeminence of environmental extremism became more pronounced, political and economic factors impacting the industry were on the increase, and additional attempts were made to place heavier taxes on mining and mineral operations. Let's face it, we are (still) in trouble—particularly the small miners and prospectors (who should probably be placed on the list of endangered species).

On the good side, increases in precious metals prices throughout the year spurred a boom in gold and silver exploration, development and production. Average increases in a few base metal prices (particularly copper) improved the profitability of some mining and smelting operations, and a gradual decline in the value of the U.S. dollar on foreign currency exchanges helped some sectors of the industry to become more competitive in the foreign and domestic markets. In addition, cost-control measures initiated by many mining companies a few years ago finally began to pay off.

On the down side, a number of mining operations were forced to close because of lower average prices and the fact that certain foreign mineral commodities were flooding the markets, while domestic (U.S.) mining costs continued to increase. Many of the operational cost increases were directly attributable to unrealistic environmental controls, higher taxes, increased bonding and permitting costs, and the time and money lost through overregulation of the industry.

There were several paradoxes evident in 1987: (1) while the

mining industry continued its efforts to become more efficient, cost-effective and productive, the negative impact of political, environmental and other economic factors tended to accomplish the opposite effect; (2) while Congress expressed a desire to implement policies that would encourage domestic minerals development and production, prohibitive environmental and regulatory policies were instituted that effectively made this objective much more difficult to achieve, and (3) while our national leaders were horrified by the country's continuing foreign trade deficits and increasing dependency upon foreign mineral imports, they have refused to implement any policies that would help mining to become more competitive.

On the positive side, the U.S. land management agencies adopted some policies that will encourage minerals development on the public domain lands. At the same time, however, they are enforcing stricter requirements for environmental studies, reclamation, bonding, permitting, and plans of mining operations. In addition, whenever a land management agency actively encourages and approves mining operations, then the environmental groups launch an all-out effort to shut them down.

Also on the up side, most federal agencies are beginning to express a willingness to work with (instead of hinder) miners and the minerals industries. On the down side, a number of state and local agencies and most environmental groups are becoming more intractable in their opposition to mining (and miners) in any form.

As can be seen, there isn't much consistency in the nation's policies toward the mining industry (or most other basic industries, for that matter). Then we also have the separate matter of the individual state and local government policies, many of which are even more inconsistent than the federal policies. Since the Supreme Court's decision in the Granite Rock case, a door has been opened for an ever-expanding control of mining operations by a plethora of state and local agencies. This is *not* a good development, and it definitely does not bode well for the mining industry in the future.

The recent drop in the stock market, plus the attendant uncertainties in the world's financial community, produced a negative impact on a large number of mining companies and operations, and the ripple effect from this situation will probably affect some sectors of the mining industry for some time. On the up side, however, uncertainties in the financial markets should exert upward pressures on mineral commodity prices—especially for the precious metals.

If this article appears to be more of a wrap-down instead of a wrap-up, it is only because these realities must be faced sooner or later (mostly sooner). If viewed through rose-colored glasses, the events in 1987 might appear to be a little more encouraging. When viewed in the light of reality, however, this past year has produced more negative effects than positive results. We can always hope that the next year will bring about some positive changes in the situation but, unfortunately, it probably won't unless we bring them about ourselves.

The handwriting is on the wall, if we will take the time to read it.

8
BLM Reclamation Bonding and the Small Miner

Author's Note: At the time of this writing, Congress was in the process of preparing legislation that would require the U.S. Bureau of Land Management (BLM) to enforce the posting of reclamation bonds for most mining operations on the public lands being managed by that agency. The following information provides an analysis of the potential impacts such legislation would have upon the nation's small miners and prospectors as well as the subsequent impact upon the exploration and development of new mineral deposits in the U.S.

THE U.S. MINING INDUSTRY has been subjected to an ever-increasing number of stringent environmental regulations in recent years, and it has now become the most overregulated mining sector on the planet. This plethora of rules, regulations and restrictions has been a major factor in the increased mining costs and time delays currently being experienced by the industry.

Unfortunately, most of these new regulations are phrased and applied in such a manner that they make little or no distinction between the major multinational mining conglomerates and the individual small-scale miners and prospectors. In other words, we must go through the same permitting, bonding and regulatory procedures that are required for major mining operations. Since this also applies to federal, state and local regulation, it has become an almost overwhelming obstacle to exploration and development of mineral deposits by small-scale miners and prospectors. Imposition of an additional reclamation bonding requirement on small mining operations would definitely produce a corresponding reduction in the amount of mining activity generated by the individual miner and prospector.

The impact of reclamation bonding requirements by certain other regulatory agencies (i.e., U.S. Forest Service, state, and county) has already resulted in an overall decrease in mining activity in some locations. The individual miners have begun to move out of certain states, counties and Forest Service districts, largely because they cannot meet the same reclamation and bonding requirements imposed on the major mining companies. This is producing a further constriction in the overall area in which small miners can, or will, continue to operate.

The current trend toward environmental extremism also plays a very important role in determining the small miner's ability to effectively compete in the modern mining business. Reclamation and operational requirements are increasingly based upon the restoral of mined lands to their original state, as opposed to the elimination of unnecessary or undue degradation of the environment. Since most reclamation bonds are based on the required degree of restoration, attempting to restore mined lands to their original state often produces extremely high bonding requirements. So depending upon the relative emphasis on environmental concerns in each particular area, the bonding and reclamation requirements can effectively close the small miner down.

Even if the larger bonds can be posted by the individual, the added costs involved drastically reduce the potential profitability of many small-scale mining operations. As a result, a large number of the smaller mineral deposits will not be mined—mainly because most of the larger mining companies will not touch them and the small miners are unable to meet the higher cost of operations.

In addition, there have already been a number of cases in which the reclamation bonds and requirements have been used to prohibit or severely restrict the development of certain mineral deposits by reducing and/or eliminating the potential for profit from the mining operations. Many of the county ordinances have been utilized in this manner as a tactic for effectively shutting down mining activity in some areas. In a few cases, reclamation bonds and permits have been required by county, state *and* federal agencies for the same mining operation. In several instances, even the larger mining companies have been forced to pull out of the district. It is obvious that most small mine operators would not have any chance of operating under these conditions.

Another sore point in this regard is the "guaranteed" mining rights to mining claims located in designated wilderness areas, national parks and/or wilderness study areas. The stringent operational, environmental and bonding requirements in these areas have caused the abandonment of otherwise viable mining operations, which means that these mineral deposits can never be mined under existing regulations. The small miners and prospectors cannot even consider any mining activities in these districts, and most major mining companies will not make the attempt unless they have located a world-class mineral deposit. Since exploration activities are also severely restricted, it is unlikely that any significant new mineral discoveries will be made (or proven) in these areas.

The following list summarizes some of the potential adverse impacts of additional reclamation bonding on the small-scale miner and prospector.

1. Most individual miners and prospectors cannot obtain bonds from the bonding companies or financial institutions. This means that they must post cash bonds from their own assets, and many part-time miners do not have this extra capital.

2. Most cash bonds posted by individuals "freeze" significant amounts of capital that would be utilized for mine development.

3. The cash value of reclamation bonds and requirements may be prohibitive if the reclamation requirements are set at unreasonable levels.

4. Increasing costs associated with permitting, bonding and reclamation requirements will cause a net decrease in mining activity by small miners.

5. Strict reclamation requirements and bonding will preclude the development of small and low-grade mineral deposits.

6. Additional capital costs will reduce the exploration for, and development of, new mineral deposits in the U.S.

7. The posting of duplicate bonds for the same mining operation with several regulatory agencies is unnecessary and unreasonable, and this tactic has already been used to prohibit mining operations rather than as a means to regulate industry activity.

A recently submitted GAO report to Congress states that reclamation bonding by the BLM will have little, if any, adverse impact on the nation's mining industry. This report is obviously referring to a general response by the mining industry to a proposal for *reasonable* bonding and reclamation requirements. If this were indeed possible, and the bonding amounts were based upon each individual case being considered, then reclamation bonding would probably not place an undue burden on the industry as a whole. However, from the small-scale miner's and prospector's viewpoint, any additional reclamation bonding would most certainly produce a negative impact on their mining and exploration activities, especially when considering the current stringent environmental standards. Mineral deposits have already been abandoned by potential mine operators and mineral exploration activity has been sharply curtailed wherever prohibitive bonding and reclamation requirements are the norm. Because of this, it should be pointed out that many of the GAO report's conclusions are erroneous.

It seems certain that additional blanket reclamation bonding for all types of mining activities on the public lands will act to discourage exploration and development of new mineral deposits by the nation's small-scale miners and prospectors. It is our contention that the current requirements for reclamation plans and compliance by mine operators, when adequately enforced under existing regulations, is more than sufficient to prevent any unnecessary and undue degradation of the environment. Although bonding for large surface disturbances might be required in many instances, the application of the same or similar bonding requirements for the reclamation of minor surface disturbances is not necessary, and it would place an additional burden on small-scale miners and prospectors.

9
Miners Are a Threatened and Endangered Species

MINERS AND THE MINING INDUSTRY have been a major target for the media, politicians, environmentalists and anti-mining radicals in 1989. We have been under constant attack on the federal, state and local level all year, and the outlook for 1990 is much the same—if not worse. Miners, prospectors and the mining industry must become better organized and much more effective if we are to combat these attempts to put us out of business. Our very survival as individual miners and prospectors is at stake.

We are currently faced with a number of critical issues, including: the proposed changes to, or elimination of, the federal mining laws; current and future wilderness designations and other types of land withdrawals; new regulatory proposals by federal and state agencies; restrictive county and local ordinances, and ongoing anti-mining activities by environmental radicals. Mining's public image continues to be misrepresented and distorted by the media, extremists and special interest groups. The ultimate objective of these attacks on mining has become quite apparent: we are slated for extinction.

As a rather explicit analogy, the U.S. mining industry is now living in the equivalent to the Paleocene Epoch (early Tertiary), during which the earth saw the extinction of dinosaurs and ammonites and the development of flowering plants. All we have to do now is step into the tar pits so our current development is preserved intact for examination in the future. The alternative, of course, is to fight back and adapt to our new environment so that we may survive.

And we are not alone. Miners, loggers, oil producers, manufacturers, energy developers, farmers, ranchers, hunters, recreationists, rockhounds, four-wheelers, motorcyclists, senior citizens, the handicapped and others are now on the radical protectionists' hit list. It seems that no one is to be allowed to utilize and enjoy the nation's great outdoors except, of course, for the environmental elitists and radicals.

Environmental activists are engaged in an all-out campaign to lock up as much of the nation's natural resources as possible, abolish the 1872 Mining Law, halt timber sales and harvests, reserve water rights for non-use, restrict industrial and development activity, and eliminate multiple uses of our public lands. Reasonable compromise has been discarded, as these self-appointed saviors of mankind pursue their utopian quest to make this country (and ultimately the rest of the planet) into one huge, nonproductive national park.

In "Common Sense Versus Emotion," Ta M. Li, president of the Northwest Mining Association, recently said:

> Our industry has learned, once again, that logic, good sense and the critical nature of domestic mineral production are not enough to combat the thrust of the preservationists. Anti-mining groups, public land non-use proponents, and anti-business sentiment have increased since the last national elections, beginning with a several-hundred-item list of demands delivered to President Bush by preservation groups. Reading the factual testimony of industry representatives at federal hearings and the emotional demands of the opposition makes one realize that the playing field is still tilted in favor of emotion....
>
> [W]e cannot expect miracles in educating Congress, federal agencies or the public. There is a serious lack of knowledge about how critical domestic production is to our economy and to our security. We are different than the oil industry in that a supply shortage is not immediately recognized by the consumer and, therefore, requires planning. As Senator McClure noted in his recent Idaho wilderness bill proposal, our new mineral discoveries are hidden from view, so that keeping lands open for minerals is a difficult job. Unfortunately, the need for mineral discoveries seems equally hidden from view.

In the July *Mining Magazine* (London), editor Tony Brewis pointed out that the general perception of the (mining) industry has not been good: the public tends to think of everything negative about it, such as scarred landscapes, windblown dust, smoke emissions and so on. Unfortunately, the industry does not do a good job of explaining its achievements, whereas the environmental activist groups, in contrast, are masterful users of the media and politicians have seized upon protection of the environment as a worthy vote-winning topic. On all counts, Brewis says, positive publicity for the minerals industry is needed *now!*

The same editorial quoted Dr. Peter Hackett, principal of the Camborne School of Mines and president of the Institution of Mining and Metallurgy, as saying the extractive industries are generally perceived by the public as being "dark, dangerous, dirty, socially unattractive and almost irrelevant to our way of life." The editorial went on to point out that if the minerals industry is not to have to submit to unwelcome legislation, then its members need to take a positive, politically active role in helping to form the laws and the public opinions that call for them to be drafted.

So, it is obvious that miners' problems are not minor problems, nor are they necessarily of the home-grown variety: the excessive protectionist mania has spread like an epidemic around the globe.

The radical environmentalists' objectives are by no means limited to the permitted use and designation of wilderness areas and other types of land withdrawals. They are also attacking any potential development activity on nonsensitive lands, both public and private. Strict regulation, stringent environmental standards, civil litigation, special use permits, bonding costs, licensing requirements, political pressure, and public opinion are quite often used as a means for preventing or discouraging new development activity rather than as a means for regulating it. Costly and time-consuming Environmental Impact Studies are the order of the day, often preventing or precluding even the simplest development project by negating its economic feasibility.

As an increasing number of our political leaders jump onto the environmental bandwagon, rationality and logic are being abandoned. The courts, having already firmly established our propensity for perpetual litigation and overregulation, continue to churn out biased rulings that serve to restrict the access to, and availability of, our own natural resources. For the first time in U.S. history, the public has become obsessed with the notion that productive enterprise and a solid work ethic are no longer relevant to their daily lives.

As with most other popular movements throughout our history, the legitimate and commendable efforts to conserve and protect our environment have evolved into a classic example of overkill. The current problems do not lie in the original objectives and accomplishments of the conservation movement—they lie in the radical extremism into which the movement has evolved. Many of the well-intentioned and highly principled members of the environmental organizations are now being deliberately misled by the use of erroneous information, biased views, examples of past excesses, and the continuous repetition of outdated philosophies and catchwords.

We can no longer afford to sit idly by and attempt to avoid personal involvement while this process reaches its logical conclusion. It is time to prepare an Environmental Impact Statement, if you will, that addresses the dangers inherent in environmental extremism. The excessive environmental protection mania sweeping the nation must be faced head-on and brought to a halt. If we cannot stop this insanity before it progresses much further, then we might as well step into the tar pits alongside the dinosaurs and other extinct species.

10
Wilderness Update: How Much Is Enough?

Editor's Note: This article, which appeared in the April 1991 issue of the *California Mining Journal*, reported on the status of wilderness preservation in the United States and the effects of excessive wilderness withdrawals on natural resources development. It was the third report on this topic over a period of several years, appearing annually in the *CMJ* beginning in 1989.

MOST OF US AGREE THAT the conservation and preservation of certain natural and sensitive areas in the United States, by designating them as protected wilderness, is both a desirable and necessary objective. It is also absolutely essential that we drastically reduce the harmful pollutants that our modern civilization produces and subsequently discharges into the environment. It is *not* necessary, however, to completely halt the development of our natural resources and deliberately slow the progress of the nation in order to achieve these objectives.

As with most worthwhile grassroots efforts to correct real problems and accomplish necessary changes, the environmental movement has evolved from its original rational approach to a definite trend toward extremes. The Wilderness System has grown to massive proportions and the efforts to make huge additional tracts of the nation's lands off-limits to any type of productive use are continuing unabated. We are locking up more and more of the public lands because of emotional appeals based upon invalid reasoning and general perceptions.

Viewing Wilderness in the Proper Context

There is a definite bias towards removing access to our country's major natural resource areas that is not readily apparent in information published by the media. We are concerned here with the most beneficial public and economic uses or withdrawals of the public domain lands in the U.S. and where these lands are located. Therefore, any figures and/or percentages that are based upon the total of all the lands in certain states or the nation overall can be deliberately misleading. It is necessary to consider the country's, and each particular state's, overall land status in order to place wilderness in its proper perspective.

The vast majority of our nation's public domain lands are located in the western U.S., which is also our most productive natural resource area. This is the region that is being targeted for almost all of the more massive wilderness withdrawals. In addition, most of this region's available natural resources and best multiple use lands are situated in mountainous and hilly terrain, and the bulk of the current and proposed withdrawals specifically cover this type of terrain.

At present, over 95% of the total acreage in the National Wilderness Preservation System is located in the 11 western states and Alaska and now covers well over 13% of the total public domain lands in these 12 states. Data for the nation overall appears in the table below:

SUMMARY OF U.S. LAND STATUS	
Total land area/U.S.	2,271,343,360 acres
Total acres/private ownership (68.12%)	1,547,277,189 acres
Approx. inland water area in the nation	50,867,840 acres
Total Federal lands in the nation (31.87%)	724,066,170 acres
Acquired Federal lands total (2.77%)	63,089,515 acres
Total public domain lands in U.S. (29.1%)	660,976,655 acres
Current land in designated wilderness	94,984,898 acres
Wilderness as 9% of total Federal lands	13.2%
Wilderness studies/withdrawals/proposals	33.93%

The Current Wilderness Picture

According to data from the U.S. Department of the Interior, several additional wilderness bills were approved by Congress in 1990 — including about 2.3 million acres of new wilderness in Arizona and several smaller additions in other states. This means that the wilderness system has expanded to almost 95 million acres, and it now covers about 4.18% of the total U.S. land area. Overall designated wilderness now covers 13.12% of all the federal lands in the nation and a higher percentage of the public domain lands in the western states. And, this is just the tip of the iceberg.

Federal agencies alone have already recommended about 20 million additional acres as being suitable for wilderness designation, and the agencies are still conducting "studies" on about 130 million more acres for possible inclusion in the system. The BLM recently recommended about 1.9 million acres for Nevada wilderness, and the final BLM wilderness recommendations for several other states are still to be finalized in the next couple of years. In addition, a number of massive wilderness proposals are being advanced by various environmental protection groups and several politicians in Congress.

The current status of the U.S. wilderness system, broken down by the agencies involved, is presented in the table below.

THE NATIONAL WILDERNESS PRESERVATION SYSTEM as of January 1, 1991			
Agency	No. Areas	Federal Acreage	
National Park Service	42	39,075,415 acres	41.1%
U.S. Forest Service	380	33,609,661 acres	35.4%
U.S. Fish & Wildlife Service	75	20,688,827 acres	21.8%
U.S. BLM	66	1,610,995 acres	1.7%
Total Wilderness*	**546**	**94,984,898 acres**	

* Detailed breakdowns of each wilderness area within specific states and by managing agency can be found in the Annual Wilderness Report to Congress and, because of the sheer volume of information involved, are not listed here.

It should be noted here that these figures only represent the current wilderness withdrawals, and the total of proposals in progress listed immediately above the table, by federal and management agencies. The total of withdrawals, proposals and study areas, both de jure and de facto, now covers roughly 34% of the country's total federal lands. The data does *not* include the numerous other types of land withdrawals, such as: Wild and Scenic Rivers, National Monuments, National Wildlife Refuges, National Recreation Areas, National Parks, or other types of land withdrawals by the Department of Defense, Bureau of Reclamation, Corps of Engineers, Department of Energy and several other agencies.

This, briefly, is the current wilderness picture, and it raises an obvious question; just how much wilderness is enough?

The multiple use concept and balanced natural resource management programs currently being implemented by our two primary federal land management agencies, the U.S. Forest Service and the Bureau of Land Management, are presently designed to achieve almost the same objectives as those being propounded by the most vocal wilderness advocates. Both federal agencies are actively seeking the elimination of undue degradation of our environment, and they are also promoting programs that will enhance the environment.

It is becoming increasingly evident that the most effective means of preserving and protecting our nation's public domain lands is through the application of proper land management policies, and most of these policies are already in effect. Why is it, then, that we cannot continue to pursue these worthwhile objectives in a calm and reasonable manner and are instead resorting to the unrealistic and unreasonable extremes?

The attention of the general public is currently being focused by the news media and preservationist groups upon the wilderness withdrawals and additional stringent overregulation as the only means whereby we can "protect" the environment. Is this really necessary, or are certain special interest groups using public sentiment in order to gain their own objectives?

11
Mining Issues in 1993: What Can We Expect?

MINING WAS AGAIN CHOSEN as the preferred whipping boy by the news media, environmental radicals and anti-mining politicians during most of 1992. Miners were constantly being depicted as dark, dangerous, dirty, distasteful and destructive by the media and those opposed to mining. A massive political and media campaign was mounted by most of the major environmental groups, with the ultimate objective of either severely restricting or completely shutting down most of the nation's natural resource industries—especially mining and logging, with ranching running a close third. Unfortunately, these efforts were quite successful in increasing the public perception that the country is being raped and ruined by uncontrolled "development interests."

This trend towards the environmental extremism that has characterized the past two decades reached a record high last year, and environmental issues now dominate the national agenda. Despite the recent open acknowledgment by the federal government that unnecessary overregulation is stifling the American economy, the bureaucracy continued to pump out an overwhelming number of new and costly laws, rules and regulations. The current preoccupation with environmental issues has accelerated this process to the point where it is seriously impacting every U.S. citizen except the members of Congress, because they continue to exempt themselves from the laws and regulations they impose on the rest of us.

In Congress, the number and critical nature of mining issues increased substantially in 1992. Attacks on the 1872 Mining Law, as amended, and demands for excessively stringent environmental controls were aggressively pushed by anti-mining politicians from the session's opening bell right up to the final adjournment. And, in a behind-the-scenes political trade-off between Congress and the Administration, the government decided to sacrifice U.S. mineral exploration by imposing the new "cash only" mining claim holding fee, with full knowledge of the fact that the fee would devastate most individual miners and smaller firms, would act to discourage mineral exploration and development, and would not even come close to producing the projected revenues to the U.S. Treasury.

Natural resource industries (as well as all other producing business enterprises) in the United States are still being overwhelmed by an ever-expanding bureaucracy, stringent and inflexible laws, short-sighted policies, escalating costs, unlimited liability, unbelievable overregulation, perpetual litigation, and a totally undeserved anti-development attitude held by certain politicians and members of the public. Mine planning has become an absolute exercise in futility because the laws, rules and regulations continue to change so rapidly that it is impossible to keep up with them, effectively precluding any reasonably accurate estimate of the future business environment for planning purposes.

Although government has consistently shown that it is totally incapable of managing itself, it persists in the belief that it can better manage all businesses and each individual's personal life. And, even in an election year when the nation should have had some breathing space, this trend toward total government control of everything and everybody continued at an accelerated pace. As a somewhat incredible example of totally irrational thinking, federal administrators and the U.S. Congress actually think they can correct the country's problems by enacting *more* restrictive laws and formulating even *more* stringent regulations. The real problem here is rather obvious.

This situation does not bode well for the future of mining in the U.S., especially when considering the results of the recent election. If any business is going to be hit with more unreasonable legislation and regulation, mining is automatically moved to the top of the list. Many politicians at the federal, state and local levels want to put mining out of business, with the full support of most environmental groups. The phrase "Mine Free by '93" was not coined by radical environmentalists as a joke—they fully intend to achieve this objective. So, what can miners expect to be faced with during 1993?

Mining Law Reform in Congress

Of the two major mining law "reform" proposals in Congress (Rep. Nick Rahall's and Sen. Dale Bumpers' "substitute" bills) that were introduced in 1992, both are expected to reappear in the 1993 session—possibly with some modification and/or revision. It is expected that Sen. Bumpers' Trojan-horse substitute bill might be extensively modified prior to reintroduction because of his failure to fool members of the U.S. Senate last year. In any event, there will be further behind-the-scenes coordination between Rep. George Miller in the House and Sen. Bumpers in the Senate—with assistance from radical environmentalists in closed-door meetings—to railroad a destructive mining law bill through Congress and get it to the president's desk for signature.

Considering the extremist environmental philosophy expounded by the new vice president, Al Gore, there appears to be a strong likelihood that the new president, Bill Clinton, would sign the measure if it manages to make its way through Congress. It still appears likely that Rahall's monstrosity (most probably pushed by Rep. George Miller and his cronies) will be approved by the House of Representatives, so opposition to these attempts to abolish the 1872 Mining Law, as amended, should be concentrated on members of the U.S. Senate. All is not gloom and doom, however, since there is an excellent opportunity to stop this insanity in the Senate if miners and the mining industry can get their act together.

Environmental Legislation in Congress

The tendency towards overkill in environmental matters is expected to increase substantially, particularly when considering the outcome of the recent elections—and, as a result, it is expected that Congress will be considering a plethora of new (and resurrected) proposals to enact even more insane and extreme environmental laws and standards. Among these will be the further review and overhaul of the Resource Conservation and Recovery Act (RCRA), the Endangered Species Act (ESA), consideration of additional water quality and air quality amendments, and special environmental legislation directed towards specific industries and problems (both real and perceived).

There is another "up jumped the devil" expectation in the environmental arena for 1993. Quite a few excessively stringent and damaging pieces of environmental legislation were ramrodded through Congress during the past 12 years, only to meet a presidential veto during the Reagan-Bush era. At the time, there was insufficient congressional support to override the vetoes. With the new Democrat administration, however, the hope that these legislative failures can be passed has increased significantly. As a result, some members of Congress are frantically scurrying about in preparation for the resurrection of these proposals. So, it is expected that quite a few of the legislative monsters of the past will arise from the dead and find new life during this (and succeeding) sessions of Congress.

It is highly likely that quite a few onerous new environmental laws will be enacted by Congress in 1993, especially when considering the ongoing national mania over environmental matters, the intense lobbying pressures exerted by protectionist groups, and the apparent "save the planet" mentality of the new administration. Some of these measures will have a tremendous impact on business, local governments and the national economy by imposing huge additional costs (for very little benefit) as they are implemented, as well as placing ill-conceived and onerous burdens on all producing enterprises to achieve "compliance." If the current extremism in these matters is allowed to continue unopposed, environmental laws alone will eventually be sufficient to shut America down.

The Mining Claim Patent Issue

This is again the most likely mining issue to be changed (or eliminated) this year, but it will most likely appear as part of a larger, much more damaging piece of legislation. The mining claim patent process serves as a lightning rod that attracts most of the attacks from the media, environmental groups, many politicians and the general public. This single issue provides a tremendous boost to the efforts exerted by anti-mining groups because it gains them support for demanding that the 1872 Mining Law be abolished. The primary reason for the defeat of Sen. Harry Reid's mining patent amendment in the FY1993 Interior Appropriations Act was the unwillingness of mining's opponents to allow a reasonable solution to the matter, because they fully intend to misuse and abuse the issue in 1993 as a means of gathering support for abolishing the entire U.S. Mining Law.

Patenting has become a major public perception issue because, even though the truth is widely known, the opposition forces and the media persist in the constant repetition of lies and half-truths in news items, environmental publications, consumer magazines, and public hearings. As a result, most of the general public is still firmly convinced that miners are ripping off the taxpayer by purchasing land "for as little as $2.50 per acre" and then selling the land to developers for millions of dollars in profits. The truth that it actually costs thousands of dollars per acre plus other stringent requirements (recently substantiated by a BLM study and news release) is now irrelevant, since it is almost never published and the lie has become firmly entrenched in most people's minds.

As a consequence of this propaganda campaign, miners have been forced into supporting changes in mining patent law in an effort to defuse an erroneous perception of the matter. It is expected, again, that a mining-initiated or mining-supported bill to amend the mining patent law will be introduced this year. The most likely amendment proposals will be to: (1) raise the per-acre patent fee to the full market value of the property; and/or (2) provide for the patenting of the mineral estate only, with the surface estate to revert back to the public domain after mining is done.

There is no doubt that this single issue will be at the forefront of most major attacks on the mining law in 1993. There will certainly be several other legislative proposals specifically targeting the elimination or revision of mining patent law, most likely included as part of the budget appropriations process and/or in land management legislation.

Mining Claim Holding Fee Revisited

The $100 mining claim holding fee per year imposed during the 1992 session will definitely be revisited, but in a negative manner insofar as miners are concerned. The idea that the federal government would even think of allowing any revenue-generating measure to lapse is laughable. Therefore, it is likely that one or more of the following will be implemented: (1) the $100 annual fee will be made permanent; (2) the fee (with other fees and charges) will become part of so-called mining law reform legislation, and (3) the annual fee requirement will be increased.

Although the devastating impact of this fee on small exploration and development projects was pointed out time and time again, proponents of the measure chose to ignore the facts and proceed with the fantasy that the fee would actually generate increased revenues to the Treasury. As with the punitive tax on the boat-building industry, Congress will sit back and wait until an entire sector of the U.S. economy is practically wiped out before they will even consider the possibility that they made a mistake, let alone take corrective action. There is only one safe bet on this issue, which is that Congress will not allow the fee to sunset in 1994.

Wilderness Legislation in Congress

It is an absolute certainty that there will be additional and renewed proposals in Congress (along with intense pressure from environmental groups) to enact massive new additions to the National Wilderness Preservation System. Despite the fact that designated wilderness now covers about 14.4% of all the federal land in the U.S. and nearly 4.2% of the nation's total area, protectionists will continue their efforts to expand the system as much as possible to achieve their objective of eliminating all natural resource development in America. This is also another safe issue for most politicians because of the general public's belief (following a thorough brainwashing) that wilderness is the only means to adequately protect and conserve our public domain lands.

At least some of the wilderness proposals will be approved by Congress (maybe most of them, considering the current wilderness mania) this year, but an estimate of the total number and/or acreage involved cannot be made because there are so many of them (and each proposal is subject to numerous revisions and/or additions before it is finally approved). However, the number of wilderness acres designated is expected to increase substantially over the next few years because most of the Bureau of Land Management wilderness recommendations are coming up for consideration.

More Stringent EPA Regulations

The Federal Environmental Protection Agency (EPA) is expected to continue the formulation and implementation of even more stringent and expensive (if not impossible) environmental standards and requirements. Following the agency's widely recognized tendency towards extreme overkill, the EPA has already set unbelievably low standards for the acceptable amounts of certain substances that pose risks of less than one in a million (or billion) to either plant or animal life, including many natural substances that are commonly present in much higher quantities in the natural environment. Some of the limits have been set so low that they cannot be measured accurately with the current availably technology.

Several knowledgeable state agency officials have said the EPA has been frantically preparing a multitude of new excessively stringent regulations that are slated to be foisted onto the states along with their attendant astronomical costs. The cost/benefit ratio of many EPA rules have gotten so far out of whack that recent studies estimate a cost of $1 million to $150 million per (theoretical) life saved. For that matter, the EPA has set very costly standards for a large number of chemicals for which there is absolutely no scientific evidence that the substances are harmful to human, plant or animal life.

The EPA's "fine 'em and jail 'em" enforcement policies are expected to get much worse, particularly if the agency achieves cabinet-level status. At the present time, EPA personnel do not appear to be concerned at all with regular compliance as long as they can send the maximum number of people to prison and collect the maximum amount of fines. The existing EPA regulations, if strictly enforced, are already sufficient to shut down America's so-called smokestack industries, yet the agency persists in its insane quest to outdo Mother Nature by "purifying" the natural environment.

More Threatened and Endangered Species

It is expected that federal and state Fish and Wildlife agencies will be accelerating the listing of more threatened and endangered species to the thousands already listed, with the consequent additional unrealistic, excessive and expensive restrictions being placed on natural resource development—if not the total elimination thereof. A major expansion of these listings is right at the top of the agenda for most environmental organizations because it quite effectively brings most development activities in many areas of the nation to a complete halt.

This is a very important "surrogate issue" for anti-everything radicals because they can utilize the ESA to achieve a de facto closure of massive tracts of land (both public and private) throughout the U.S. without having to obtain congressional approval for numerous land withdrawals. This is also another public perception issue: the general public has been misled and misinformed for so long that they actually believe it is necessary to stop natural resource activities in order to protect all those furry, cute and cuddly critters out there from the greedy and insensitive pro-development interests. In actual fact, it has been repeatedly proven that innovative resource, wildlife and land management policies actually increase wildlife populations and enhance wildlife habitats, whereas Mother Nature continues to implement the basic "survival of the fittest" program.

Unfortunately, this biased public perception is not going to change in the foreseeable future, either; it has become firmly entrenched in most people's minds. As a result, it appears to be somewhat unlikely that efforts to make reasonable (and necessary) changes in the Endangered Species Act will be successful. For that matter, it seems much more likely that the ESA will be amended in such a manner as to make it an even greater disaster.

New BLM Mining Regulations

At the present time, the Bureau of Land Management has finalized the new regulations which require financial guarantees for mining activities involving a surface disturbance of five acres or less and bonding for mineral exploration and development projects that disturb over five acres. The new requirements would have been implemented as early as October 1992 if not for the president's moratorium on the issuance of new regulations. However, BLM officials have applied for a waiver from the Office of Management and Budget (OMB), and as soon as this is approved the regulations will be published in the Federal Register the next day. It is expected that some form of financial guarantee or bond (surety) will be required for all mining activities that exceed casual use, and the regulations may take effect as early as January 1993.

In addition, the BLM is currently in the process of rewriting the 43 CFR 3809 regulations, particularly in reference to the 5-acre rule for notice-level operations, and the agency is also expected to revise several other provisions, including: requiring approval of a mining plan of operations for small mining activity; defining the term "unnecessary and undue degradation"; specifying prohibited acts, with provisions for civil and criminal penalties and enforcement; specifying additional environmental and reclamation requirements, and clarifying activity allowed under "casual use." It is expected that these new regulations will become effective fairly early in the year.

The BLM also issued a new rulemaking proposal to implement

the annual $100 mining claim holding fee requirement in lieu of the federal annual assessment work requirement in early December. Following a short comment period, these new regulations are expected to be finalized and implemented on a priority basis—most probably by late February to early March. This rulemaking process might prove to be interesting due to several rather notable defects in the empowering legislation (FY1993 Interior Appropriations Act).

The BLM is also currently finalizing its new rules on occupancy of mining claims, and these new regulations are expected to take effect in early 1993. Overall, the new requirements of these new BLM regulations will most certainly have a significant impact on miners (particularly on small-scale miners and mineral exploration firms), especially because of the combined effect of implementing so many new regulatory requirements over a fairly short time period. This rulemaking process constitutes a *major* mining issue in 1993 because it has the potential for removing some (or all) of the basic protections for small-scale miners and companies against the current national tendency towards bureaucratic overkill.

More Land Restrictions and Withdrawals

With the considerable efforts being made by environmental organizations, there will undoubtedly be an increasing number of designations of Areas of Critical Environmental Concern (ACECs), National Conservation Areas (NCAs), National Recreation Areas (NRAs), and other restricted land management areas by Congress and federal agencies. These classifications are especially attractive to most anti-mining groups because they are usually quite successful in eliminating or severely restricting all types of mining activities in special-use areas. The number and type of these special-use land categories have been proliferating over the past few years and their number is expected to increase significantly over the next couple of years. Land withdrawals by the Department of Defense, Bureau of Reclamation, Corps of Engineers and other government entities are also expected to increase.

Stringent State Environmental Laws

Also continuing the trend toward extreme overkill in environmental matters, many state legislatures will be considering a number of proposals to enact even more stringent environmental statutes and state agencies will be formulating excessively stringent environmental regulations. A major portion of this insanity is a direct result of the Federal EPA cramming unreasonable and unjustifiable trace-element standards down the throats of the states (with the threat that they will enforce the standards if the states do not). The horrendous costs associated with these sometimes unattainable new standards are now largely passed on to local communities (cities, towns and counties), which are nearing the point where they can no longer fund these "purification" programs without destroying their economic base.

Most of these proposals will be subjected to intense lobbying efforts by environmentalists and, again recognizing the mistaken public perception of these issues, many of the measures will be approved.

Increased User and Permit Fees

Congress, state legislatures, state agencies and federal agencies (as well as local agencies in some instances) will continue to propose a plethora of new user and permit fees for all types of mining activities, in addition to the constant requests for permission to increase fees for existing permits and agency "services." This has become a very popular means for getting mining and other types of businesses to pay for an expanded bureaucracy, increased regulation, more inspections, and much tougher enforcement. In other words, you are required to pay more of your hard-earned money to help the bureaucracy make it much more difficult for you to stay in business.

More Lawsuits against Mining

It is absolutely certain that many of the environmental and protectionist groups will continue to misuse and abuse the civil and administrative court systems by filing numerous frivolous lawsuits to delay or halt proposed mining projects and most other natural resource development activity. Any reasonable decision and/or policy to be implemented by federal and state agencies will be a prime target for these groups, because they can quite effectively shut down almost any potential natural resource development activity by tying up huge tracts of land in the court system for many years at a time. In addition, both civil suits and administrative appeals will be utilized by anti-everything radicals to halt specific projects or, at the very least, create costly and time-consuming delays in the hopes that the economic feasibility of a project might be destroyed.

More Public Attacks on Mining

It should come as no surprise, and it can be stated with absolute certainty, that there will be frequent attacks on mining by certain politicians, anti-development groups, and environmental radicals. These assaults will be part of the strategy employed by extremist elements in the environmental movement to eliminate all types of natural resource development or use everywhere, for any reason, for all time. The attacks on mining will be fully supported (and often initiated) by television network programs, articles in major national magazines, reports and news items in newspapers, and radio talk show hosts.

Biased reporting and repetition of untrue information will be intentionally utilized in an organized propaganda campaign to gain more public support for abolishing the mining laws and placing severe restrictions on all types of public land use and natural resource development. If the mining industry cannot organize and implement an effective counter campaign to address this media propaganda, miners will continue to lose the battle for survival in both the political and public arenas.

Political Pressure

The current preoccupation with environmental issues makes it certain that both federal and state legislators will be under constant pressure to support purported "pro-environment" issues and oppose any alleged "anti-environment" issues. The threat of active political opposition (especially in the press) by any powerful special interest group, as well as the dangling carrot of promised support, will undoubtedly have a significant influence on the actions taken (or not taken) by many of the people in the political arena.

The World Economy

Economic activity throughout the world in 1993, especially in the production of finished goods and the construction industry, will have a considerable impact on mining. Lower materials consumption by the industrial sector translates into lower demand and lower mineral prices. Higher unemployment, reduced income levels, and slow economic activity decrease the availability of investment capital and the money available for discretionary spending (which would translate into lower demand for precious metals in jewelry and investment products). The main mining issue here is the potential for lower production, closure of marginal mines, and mines laying off employees to cut costs. This occurred over most of the past three years, is continuing at present, and could get much worse if the world economy remains in a recessionary state or goes through a period of no, or slow, growth.

Mining Industry Perceptions

This has become a major issue that directly impacts the business aspects of mining. As mentioned previously, U.S. business planning over the long term has become extremely difficult (sometimes impossible) because factors affecting each individual business are changing so rapidly there is no way they can be accurately forecast. The situation is much worse in the mining industry because there is a constant threat that the industry could be wiped out in one fell swoop by irresponsible legislative or regulatory actions. This places a rather dark cloud over the future of U.S. mining activity, and it is already producing negative results.

No practical businessperson will continue to invest time, effort or money in a project that can be closed down at any time, at the whim of something or somebody totally beyond the business's control. The risks are just too great. As a result, the mining industry has already begun to cut back on grassroots exploration for new mineral deposits in the U.S. and has been looking very closely at mining opportunities in other countries where mining enjoys a more favorable business climate. Several major firms have already sold their U.S. holdings and moved offshore. Numerous companies have halted their expansion and growth programs in the U.S., are expending their exploration and acquisition capital on foreign mineral properties, and are currently mining out and closing down their existing U.S. operations.

Unless miners are in some way assured that they will be able to mine the mineral deposits they discover and develop, they will continue this trend towards moving to foreign countries and/or shutting down their U.S. operations. The uncertainties presented by existing legislative and regulatory proposals is already sufficient to eventually decimate the U.S. minerals sector, even if they are never passed and/or implemented. So, the mining industry's business perceptions and risk assessments concerning the potential cumulative impact of these proposed actions comprise a critical mining issue.

12
"FederalSpeak" Dictionary

Editor's Note: This article was featured in The Bawl Mill column of the *California Mining Journal's* September 1993 issue. It is one of several satirical pieces that the author wrote for the *CMJ* over his 12-year association with the magazine.

THE RECENT PROLIFERATION of vacuous terminology utilized by the new federal administration and members of Congress has reached the point where fairly common words in the English language are no longer meaningful to the average U.S. citizen. It seems appropriate, therefore, to redefine some of the current terminology in terms consistent with up-to-date usage meaning and intent, thereby providing an opportunity for most Americans to actually understand exactly what the U.S. Government intends to do to us and our descendants in the immediate future.

The following words, "catchwords," and phrases have been redefined in both their present and future tense for additional clarity:

1. **Economic Stimulation:** taking your money and giving it to someone else to spend (primarily government).
2. **Economic Growth:** the accumulation of massive debt to sustain unproductive activity and finance expansion of government.
3. **Productivity:** the net effort versus cost expended on nonproductive enterprises minus the net loss of eliminated productive enterprises.
4. **Mining:** an historical term for a past productive activity that has been legislated and regulated out of existence in the United States (e.g., part of the net loss noted in #3 above).
5. **Miner:** an anthropological term referencing an individual who once engaged in #4 above.
6. **Logging:** see #4 above.
7. **Logger:** see #5 above.
8. **Ranching:** see #4 above.
9. **Rancher:** see #5 above.
10. **Contribution:** money and valuable goods taken from selected

individuals in order to support programs (taxation, for example) designed by those who believe they are intellectually superior to the aforementioned individuals.

11. **Fair Share:** the amount of wealth confiscated from selected businesses and individuals over and above the amount confiscated from the average citizen, the resultant revenues to be distributed to the poor masses (i.e., the "Robin Hood Principle," i.e., Socialism/Marxism).

12. **Misspeak:** (a) to orally mislead, misinform or falsify information for achieving a desired objective (i.e., propaganda); (b) to lie deliberately; or (c) to make a gross mistake through either stupidity or complete ignorance.

13. **Congress:** (a) an assembly dominated by aristocratic socialists who exempt themselves from laws imposed upon the average American citizen; (b) a legislative body intensely dedicated to reelection and the furtherance of utopian idealism; and (c) a collection of elected persons the bulk of whom have become permanently divorced from reality.

14. **Federal Administration:** the executive branch of government, which is adequately described in #13(b) and (c) above, and which strives to achieve #13(a) above.

15. **Job Creation:** expanded employment in the government and environmental sectors minus the employment eliminated in the natural resource and industrial sectors.

16. **Wilderness**: a term that describes the entire United States by the year [2025], excepting multi-level urban ghettos established to confine the human species.

17. **Endangered Species:** all life forms that currently exist on the planet, excepting Homo sapiens.

18. **Environmental Protection:** removing mankind from the earth's environment and leaving the planet subject to all natural forces without any interference (e.g., wildfires, floods, drought, hurricanes, typhoons, plagues, species extinction, etc.).

19. **Federal Budget:** an insatiable monster that consumes the national wealth; an open-ended credit line backed by the American people.

20. **Federal Deficit:** an ever-increasing mortgage on the economic future of our descendants in perpetuity.
21. **Economic Recovery:** a massive increase in the consumer debt load in order to create increased federal revenues.
22. **"A Vision for America":** a continuous expansion of government by the absorption of the private sector (i.e., state-owned enterprises replacing privately owned businesses).
23. **Natural Resources:** a term describing those resources situated in the natural environment that are to be left in a pristine, untouched state.
24. **Private Property:** a term describing real estate that is exclusively owned by the Federal Government.
25. **Public Lands:** an archaic term that will soon fall into disuse as it is replaced by the term "federal lands."

PART II

Miners vs. Anti-mining Extremism

"The major implication here is that everything touched by mining activities becomes contaminated and is no longer usable by mankind. According to this line of reasoning, the entire natural, undisturbed surface of this planet is forever polluted and therefore unusable."

13
Extremists Mount Offensive in Environmental War

THE RADICAL EXTREMISTS have risen to the top of the environmental movement, and they are now effectively controlling the direction and policies of most conservationist groups. Well-funded and articulate professionals are currently mounting a massive campaign to "protect" this nation's natural resources by any and all means available, whether we want or need their particular brand of protection or not. In reality, the nation's small miners, prospectors and the mining industry itself are engaged in a war for survival, and it's time the dangers are recognized.

The environmental battles have been fought for some time now; for so long, in fact, that most people within the minerals industries have come to regard them as a normal occurrence in our modern society. This time, however, the situation is much more serious. Having been very successful in waging and winning piecemeal attacks on the natural resources industries so far, the radical environmentalists are now mounting a coordinated, nationwide effort to consolidate their power and gain their ultimate objective: complete control of all activities on this nation's public and private lands.

To illustrate the real dangers represented by radical environmental extremism, here is a general outline of their strategy and tactics:

1. Funding: These organizations obtain financing for their activities by soliciting contributions from the general public; by sales of magazines, films and books about nature and wildlife; membership dues and club subscriptions; federal and state grants for natural science studies; holding conferences, fairs and seminars; private contributions, and other revenue-producing activities. Much of this funding is obtained through nonprofit organizations so that no taxes are paid.

2. Political: Leaders and members actively support the politicians and administrators who will support and advance their beliefs and programs. They also lobby incumbent politicians to influence their opinions and votes, and they vehemently oppose any public official that does not toe the line in their views. In addition, they provide financial and organizational support for the election of selected politicians and tell their membership whom to vote for and whom to oppose.

3. Administrative: They literally "pack" public hearings with vocal proponents of their cause, thereby ensuring (in most cases) that their views predominate. They propose increasingly stringent over-regulation of any and all development activities on both public and private lands—on a federal, state and local level. They are also engaged in lobbying and pressuring administrators to accede to their demands.

4. Legal: These groups retain attorneys to file suits in both civil and administrative courts in order to halt any activity to which they object. Environmental organizations sue federal, state and local agencies on *any* pretext, with the ultimate objective of obtaining control over administrative and legal procedures and policies. They also sue other companies, organizations and private individuals to achieve their own ends. In our litigation-conscious society, they are very successful in this effort.

5. Media: An extensive publicity campaign is being conducted by most of these groups and individuals to focus attention on "worst case scenarios" and examples of past excesses that have damaged, or potentially might damage, the environment. By using these examples of obvious hazards and liberally salting any current development activity with what-ifs, they have been successful in focusing television, radio and newspaper coverage on all proposed developments of our national resources. The resultant intensive scrutiny of these operations has made it extremely difficult to conduct any new development activity in an environmentally "acceptable" manner.

6. Public Opinion: A constant barrage of unfavorable publicity in the media is accompanied by a concerted effort to further influence public opinion on a state, local and national level. This objective is accomplished by radical views expressed in in-house publications (magazines, pamphlets and TV productions); by loud, vocal opposition to certain policies on local and national levels; by making erroneous and misleading statements to the press; by writing letters to newspapers and local officials, and by attempting to influence individuals through personal contact.

7. Community Activism: A concerted effort is made to organize and support any local opposition to particular issues, policies and developments. The same effort is directed toward organizing and supporting any local movement that is supportive of the environmentalist's views and objectives.

8. Regulatory: Constant pressure is exerted to add additional requirements and regulations at all levels of government. The objective here is to make most development activities much too difficult (or impossible) to accomplish by establishing unattainable standards, or to require costly and time-consuming measures that effectively discourage any progressive natural resource development. This type of activity is promoted and enhanced by efforts in all of the categories listed above.

9. Land Withdrawals: Environmental groups are constantly seeking the withdrawal of massive land areas from any type of development activity. The prime objective here is to designate additional tracts of land as wilderness, wildlife refuges, national monuments, national parks, historical districts, wild and scenic areas, special-use areas, and environmentally "sensitive" sites—primarily because almost *no* development is allowed under these land classifications. The next-best environmental objective is the wholesale designation of wilderness study areas, because they provide de facto land withdrawals for many years while also providing additional opportunities for years of litigation in the nation's judicial system. Also included in this category are lands that fall under stringent restrictions emplaced by state, county or local agencies through statutes, ordinances, land-use plans and zoning. Many of these state and local regulations effectively eliminate certain classes of land development.

This list could be expanded somewhat, but it should be sufficient to illustrate the overall strategy and tactics being used by today's environmental movement. As can be seen, this represents a powerful and dangerous combination of policies and objectives that is currently being pursued by most environmental groups. It should be stressed that this is *not* an overstatement of the situation—the situation is, in fact, much worse. Before proceeding any further, however, some background information should be noted.

The listed plans were originally formulated and adopted by the more conservative and rational members of conservationist groups in their effort to address some very serious and legitimate objectives: namely, the prevention and correction of environmental degradation that was a common occurrence in years past. These groups recognized a very real problem and set out to ensure that it was acknowledged and corrected. Largely through their efforts, the problem was eventually addressed by legal, administrative and public actions that established a comprehensive regulatory framework to prevent such abuses from occurring in the future and correct those that had already taken place. The conservationist movement *did* bring about some necessary changes in our perspective and treatment of the natural environment.

As the movement gathered momentum, however, there was a gradual shift in the ultimate direction and objectives being sought—and greed and power entered the picture. The more radical and extremist elements in the groups began to rise to the top and, since the organizational framework was well-emplaced, utilized these groups and the movement as a power base from which they could achieve their own objectives. *This* is the *problem*.

Rationality and reality have been abandoned by most environmental leaders, and radicalism and extremism have taken their place. Having very few legitimate reasons left upon which they could base their ever-increasing demands, the radicals are now resorting to scare tactics and deception to influence political and public opinion for their own purposes. By utilizing disinformation, misinformation, distortions, half-truths and bald-faced lies, these extremists are successfully deluding the media, politicians and the general public. Having already been softened up by a constant barrage of environmental horrors (both real and perceived), the average American readily accepts and believes this tripe.

Lacking both an effective public forum and any established credibility, the so-called "dirty" industries are almost hamstrung: they have no readily available means to counter this adverse publicity or the consequent public perception of their purported negative impacts on the environment. As a result, any single vocal, radical extremist can usually bring a planned natural resource development project to a halt all by himself or herself—especially if the project is located near any urban or suburban community or is in an "environmentally sensitive" area. All the radical has to do is incite a ready and willing (and primed) public. This is not only an extremely bad situation, it is insane.

And now, the various environmental groups have joined forces to form a coalition that is mounting an all-out offensive, in a major effort to win their objectives as rapidly as possible. They have stated publicly that the time is ripe for obtaining their final demands, and they are quite serious. As the saying goes, "Power corrupts, and absolute power corrupts absolutely." Now they want it all.

We can no longer sit idle and attempt to avoid personal involvement while this process reaches its logical conclusion. Our only hope for survival lies in the use of the same or similar tactics employed by the extremists (minus the deceit and lies, of course). If you think this issue will not involve you personally, think again and take a look at who is currently being targeted by the radicals: prospectors, miners, recreationists, the minerals industries, oil producers, energy companies, land developers, ranchers, farmers, rock hounds, hunters, four-wheelers, motorcyclists, loggers, manufacturers, the handicapped, senior citizens, and others: in short, almost *anyone* who might use and enjoy our country's outdoor environment.

These extremists have already succeeded in having roughly 90 million acres of the public lands designated as wilderness (an area about the size of the state of Montana), or close to 12.5% of the nation's public lands. They are currently attempting to add over 50 million more acres to the wilderness system. Senate Bill 7 (Cranston) would close most of the Southern California desert in one fell swoop (over 7 million acres), much in the same manner as they managed to virtually lock up most of Alaska. They are now targeting most of the other western public land states for additional wilderness withdrawals. Keep in mind that these acreages *do not* include other types of land withdrawals, or those that are in de facto withdrawal because of federal, state and local land use restrictions.

Additionally, the wilderness issue is ongoing; it will never be finished. As soon as the initial congressional legislation is enacted to designate wilderness areas, suits are then filed in federal courts to prevent the release of unsuitable study areas back into a multiple use classification. Litigation is also used to prevent the release of lands found unsuitable for wilderness designation during the study process. As a result, most of these lands will be in a de facto withdrawal status for many years to come. Meanwhile, environmentalists will be pressuring legislators for additional bills to create more wilderness areas and "buffer zones" around the existing designated areas.

The combined effect of land withdrawals, restricted land use classifications, impossible environmental standards, burdensome and overly restrictive regulations, and a generally hostile public will result in the eventual destruction of our natural resource industries. It will also severely restrict the opportunities for the average American to enjoy and use our nation's public lands. In addition, it will disallow many citizens from developing their privately owned lands in the manner in which they would like, and prevent some land owners from using their personal property in any way at all.

The "environmental protection" mania that is sweeping the nation must be confronted and brought to a halt. If we cannot stop this insanity before it progresses much further, we—and our country—will find ourselves in a very dangerous position. Freedom, once it is lost, is not easily regained.

14
Environmental Activists Organize to Eliminate Mining Laws

HAVE YOU HEARD the latest news from our nation's anti-mining activists? Several coalitions of the most powerful environmental groups in the United States have formally announced that they will actively support all efforts to stop and/or restrict mining activities on a national, state and local level; formulate and lobby for Congressional legislation to eliminate the 1872 Mining Law; seek the establishment of huge "ecosystem" buffer zones and special management areas around all "environmentally sensitive" land withdrawals and promote additional massive wilderness designations plus major expansions of the country's national parks, wildlife refuges, Wild and Scenic Rivers, and Special Management areas. Work on these projects is well under way, and these people are currently working full-time to achieve their stated objectives.

So, a major confrontation with the environmental extremists can no longer be avoided or put off until some future date. They have decided that the time to strike is *now*, while the movement is at the height of its power and influence, and miners are not yet effectively organized. These activist groups are well-funded, articulate, and highly effective and presently enjoy a favorable political and public image. We *must* take them seriously.

The so-called "environmental movement" has taken a new direction: By constantly emphasizing the highly publicized examples of serious environmental degradation, toxic waste and pollution—and the *very real* dangers presented by these problems—they are attempting to justify their attacks on all types of natural resource development and land use in the United States. Except, of course, their own personally approved development and land uses for their own special interests. In a mind-numbing exercise of statistical misinterpretation and biased numbers-crunching, they are attempting to prove that mining, timber, oil and agricultural interests are *all* environmental disasters.

Let's take a careful look at some of the means by which the environmentalists intend to accomplish their purposes.

A new organization, called the Minerals Policy Center (MPC), has been formed by several leaders in the environmental movement. It is based in Alexandria, Virginia, and the board of directors consists of: Stewart L. Udall (chairman), past Secretary of the Interior (1961-69); Philip M. Hocker (president), past officer in the Sierra Club; Thomas L. Kimball, honorary president of the National Wildlife Federation; J. Michael McCloskey, chairman of the Sierra Club, and Thomas A. Troyer, attorney with Caplin & Drysdale. Stated objectives: (1) *Reform* (read: abolish) the 1872 Mining Law; (2) offer *direct help* to local groups opposed to mining; (3) implement stringent regulatory controls of mining on a federal, state and local level, and (4) coordinate environmental activism throughout the U.S.

The premier issue of *Clementine* (Autumn 1988), the MPC's publication, contains some illuminating statements, several of which will be included in the following information. MPC Chairman Stewart L. Udall emphasizes the organization's policies in plain language. Excerpts from his letter state:

> Welcome to Clementine, and welcome to the Minerals Policy Center. We are a new organization, taking on an old threat—a crucial threat—to America's public lands and environmental safety. We need your support.
>
> On the public estate mining is still like a search-and-destroy mission. Mining claims and oil leases block wilderness designation in many wild, beautiful places of the California Desert and Montana's Rocky Mountain front. Mining-caused pollution has destroyed fisheries in hundreds of miles of streams like Birch Creek Wild River in Alaska. And the idiosyncrasies of the 1872 Mining Law invite squatting and further abuses on our public lands....
>
> We need not tolerate the abuses of the Mining Law and haphazard oil development, simply to satisfy our industrial urges. We believe that the problems you are facing are the strongest evidence that sweeping change is needed. Call on us...

Phil M. Hocker, president of the MPC, states in "An Introduction to the Minerals Policy Center":

> Hardrock mining and oil and gas development are responsible for serious environmental and ecological damage.... Hardrock mining creates hazardous waste at more than triple the rate of all non-mining industries combined. Wilderness areas, National Parks, key wildlife habitat areas—and pure groundwater and soil free from toxic chemicals—all fall prey to the mineral industry's claim to superior rights, a claim reinforced by biased law.

Environmental Activists Organize to Eliminate Mining Laws

To respond to this threat, the Minerals Policy Center—'MPC'—has been established in Washington to: Educate and assist grassroots citizens' groups working with environmental threats from minerals development, particularly under the 1872 Mining Law and the 1920 Mineral Leasing Act; serve as a Washington source for information on the environmental problems with those laws, and for technical data on mining methods and their impacts; lobby for reform amendments to the laws and the corresponding Federal agency regulations (the Sierra Club advocates a minerals leasing system); and to encourage improved State action to reduce mineral development impacts on State and Federal domains....

As an expert 'Service Bureau' for local groups, the Minerals Policy Center provides technical, legal and political strategy assistance to deal with mining and drilling threats to sensitive areas. It draws on examples of winning strategies used in the past, and trains local activists to deal successfully with the technical nature of mineral administration...the Center aims to unify the large amount of environmentalist concern now devoted to isolated disputes over mining around the country into a coordinated national voice.

In "Mining Law's Flaws—A Congressional Assessment," Sen. Dale Bumpers of Arkansas states his position in a reprint from the Congressional Record of Oct. 18, 1988—in no uncertain terms:

One item that is high on my personal agenda is reform of the 1872 Mining Law. I plan to request that the Energy and Natural Resources Committee conduct oversight hearings in the spring of 1989 on options for reform and to follow those hearings with proposals for changing this anachronistic law. ...The mining law provides for no royalties, lease fees, bonuses, or other payments for Federal minerals. Obviously, the 1872 Mining Law needs to be modernized to conform to current public land policy. Over the past 20 years numerous private, government and congressional studies have recommended either revising the mining law or repealing it completely.

An article by an unnamed author in *Clementine* entitled "Mines of Poison" includes the following statement:

Mining in the United States generates thirty billion tons of solid waste per year. ...Metals mining and milling alone produces over 800 million tons per year of waste classified as hazardous by EPA standards. ...Resolution of the problems of mining waste will require changes in the 1872 law authorizing rampant mining, as well as emphatic administration of the hazardous-waste laws.

Under "Membership," the publication states:

The Minerals Policy Center is a nonprofit membership organization dedicated to reforming America's mining and oil policies to protect our environment and wild lands from damage. We invite you to join us. Your membership will bring more than a tax-deductible donation to this effort. Now, your gift will be doubled by a matching grant to MPC from the American Conservation Association. By joining, you become an important part of the grassroots campaign for mining/leasing reform, and you help the Minerals Policy Center gain the right to bring lawsuits against mineral-development abuses.

The above excerpts from the MPC do not leave any doubt as to its intentions. And this is just one organization—the so-called tip of the iceberg—out of many that are expounding similar views and objectives. There are also the overwhelming number of stringent and unrealistic regulations currently being implemented by the government administrative agencies, plus unasked-for input by other departments. For example, the General Accounting Office recently submitted a self-initiated report outlining more stringent requirements for mining and public land uses. The U.S. Forest Service, the Bureau of Land Management and the Environmental Protection Agency have all issued new "final rulemaking" on natural resource development and public land issues within the past few months, and many state and local governments have recently enacted stringent mining laws and ordinances.

This "movement" is a concerted, concentrated and comprehensive effort brought about by environmental activists and extremists exercising undue influence on governmental agencies, politicians, and the general public. The results have been orchestrated by the individuals who head the nation's environmental organizations, and they are very far from being satisfied with their overwhelming successes to date.

As another example, consider the recently published *Blueprint for the Environment—Advice to the President-Elect (George Bush) from America's Environmental Community*. The publication was produced by, and with input from, 18 of the country's most powerful and influential environmental groups, and the contents address almost all of the critical environmental problems facing human civilization on the entire planet. Much of the information and concerns included in the report is very real and critical to our survival. These problems should be addressed now! However, the "blueprint for the environment" *also includes several assertions that almost all natural resource development is dangerous, damaging and distasteful.* It also recommends the president take steps to withdraw massive new areas of the public lands to "protect" them, establish buffer zones around existing withdrawals, increase stringent regulations on natural resource development, and get rid of the 1872 Mining Law. Several excerpts from the chapter entitled "Our Land" are included in the following:

...More integration is needed across boundaries (of wilderness areas, national parks, etc.), based on ecological principles. There is a pressing need to develop integrated regional resource management plans.

...We often exploit resources and divest of them too hastily at prices that do not provide a proper monetary return...

...We need forceful efforts to better protect our national park, wildlife refuge and wilderness systems from the adverse impacts of civilization encroaching upon them, and from public over-use within them. We need to expand our National Park System to embrace new additions and to reconfigure the boundaries of some park units to better protect ecosystem values.

...We must increase the emphasis on ecological management of our national forests. We should reduce timber sales on the national forests to sustainable levels, and eliminate the sale of timber below cost. A moratorium should immediately be placed on the cutting of old growth timber on Pacific Northwest forests.

...We need to increase the emphasis on ecological management and resource protection on public domain lands managed by the Bureau of Land Management, and provide better controls over the exploitation of timber, minerals and forage. We should pursue vigorously the rounding out of our National Wilderness Preservation System, with particular emphasis on wilderness designations on BLM lands.

...We need to expand greatly the protection of the remaining free-flowing stretches of our Nation's rivers, as part of our Wild and Scenic Rivers system, and intensify our efforts to work with and encourage state and local governments to take complementary actions.

...We must exert better control over mineral extraction on and off the public lands by revising the obsolete 1872 Mining Law and by better implementation of the Surface Mining Control and Reclamation Act of 1977.

A rough and simple translation of these statements would be: Make it almost impossible for our natural resource industries to operate and close off almost all of the remaining useful public domain lands to "protect" them. Abolish the 1872 Mining Law, because it stands in the way of achieving these objectives. This is the author's analysis, and it is a fairly realistic assessment of the end result if these policies are implemented.

What has happened to the efforts toward reasonable compromise on these issues? It has obviously been rejected by environmental extremists: they now want it all. Keep in mind that these actions are being initiated at the same time that almost all of the natural resource industries are making extraordinary efforts to prevent environmental degradation and are, in numerous instances, initiating programs to enhance the environment. In addition, the U.S. land management agencies are concentrating on improved enforcement of the existing protective regulations and legislation and are taking comprehensive steps to ensure adequate compliance with the numerous laws, rules and regulations.

To further compound and confuse the issues involved, these new demands are coming at a time when most of the people involved in natural resource development are just as environmentally conscious and concerned about the protection of our planet as are the environmental groups themselves. There should really be no sound reasons for major opposition on most of these issues if a reasonable, rational approach to their resolution is taken. We *can* protect the environment *without* destroying the country's ability to produce the raw materials necessary to maintain our way of life.

However, the more radical environmentalists have chosen to take a different course of action. At this time, almost all of the environmental coalitions and groups are holding meetings and seminars throughout the nation so they can effectively coordinate and train activists to successfully oppose any and all proposed mining activity and harass existing mining operations. They are instructing their members in the most effective ways to influence legislatures, the Congress, individual politicians and the general public for the express purpose of achieving their ultimate objective—total control over the use of the nation's public domain lands. In order to accomplish this objective, *they must first remove the final obstacle to their desired ends—the U.S. Mining Law.* In concert with this movement, they are also concentrating on the enactment of restrictive mining legislation on the state level and anti-mining ordinances on the local level.

Activists are also being trained in the proper ways to initiate lawsuits in the courts to halt all types of natural resource development.

This is not some projected future scenario—it is happening right now! Legislative and legal workshops are being conducted by environmental groups to provide the training necessary to enable their activists to be as efficient and effective as possible in halting mineral development. It is also likely that other types of natural resource industries are being targeted in certain areas. This isn't guesswork—these workshops and their objectives are being advertised and promoted in several publications, and a number of them have already been held.

There is no longer any room for doubt or complacency: we have to meet this new threat head-on. It is a matter of survival for miners and mining, because if we do nothing we will be eliminated. And so will the nation's natural resource industries. The environmental extremists have clearly stated their plans, policies and objectives; it's now up to us to do something about it.

15
Mineral Policy Center Supports Environmental Extremism

Author's Note: This item provides more up-to-date information concerning the topics discussed in the article entitled "Environmental Activists Organize to Eliminate Mining Laws," which appeared in the February 1989 issue of the *California Mining Journal*. The serious nature of the current efforts by environmental radicals to eliminate mining in the U.S. cannot be overemphasized: When the Sierra Club adopted the slogan "Mine Free by '93," their ultimate objectives finally became apparent. They are committed to their cause and will stop at nothing to achieve their goal.

IN ITS MOST RECENT ISSUE (Winter 1989-90) of *Clementine, The Journal of Responsible Mineral Development,* the Mineral Policy Center (MPC) in Washington, D.C., outlines the organization's proposed "Red, White 'n Blue Mining Act" to replace the 1872 Mining Laws. Not much of a surprise here: Their "draft" mining law reads much like Sen. Dale Bumpers' S. 1126, "Mining Law of 1989," introduced in Congress in June 1989 (see "An Analysis of Sen. Bumpers' S. 1126..." Jan. 1990 *CMJ*). In fact, MPC Chairman Stewart L. Udall and MPC President Philip M. Hocker both testified in favor of S. 1126 before the U.S. Senate Subcommittee on Mineral Resource Development and Production on June 7, 1989.

Udall states, in the autumn 1989 issue of *Clementine:*

Arkansas' Dale Bumpers took Mining Law reform into the U.S. Senate in June, with the introduction of S.1126, his 'Mining Law of 1989.' While the Mineral Policy Center believes this bill should have some strengthening amendments, S.1126 is an excellent start toward bringing American mining policy into the 20th Century. I testified in favor of the bill on June 7th, together with MPC President Phil Hocker.

Note that Udall said *the MPC believes this bill should have some strengthening amendments.* Bumpers' bill would have eliminated all of the small miners, dredgers, recreationists and most prospectors, along with the interdependent small equipment suppliers, mining publications and peripheral businesses. It also would have produced rather severe impacts on the larger mining companies and the industry as a whole. So, just what does the MPC mean when they refer to "strengthening amendments"?

The MPC's "Red, White 'n Blue Mining Law" would require federal, state, local and public approval before any mine development could take place, with a stipulation that the claimant "mine or get off the claim." So, if anyone objected to the mining activities proposed for any particular area then approval would be withheld, at which time the claim would be taken away from the locator. Miners could *prospect* freely just about anywhere, but they could only *mine* where they were told it was okay to do so by this discretionary "approval" process.

Exploration permits would be required and cost a reasonable fee for inspection and processing, starting at a minimum of $500 and adjustable upwards, and higher fees would be required to cover the cost of "environmental review for large projects."

Claimholders would have to pay an annual rental fee to the government for the lands covered by their claims, amounting to a "not unreasonable" $10 per acre per year. Since the law would also stipulate a 40-acre claim size (same as Bumpers' proposal), it would only cost the miner $400 per year per claim—payable in cash to the U.S. Treasury. The holder of a 10-claim block (400 acres) would only have to pay an annual rental fee of $4,000 to maintain their "privilege" to "tie up" the area covered by claims.

In addition, a fixed fee would be paid to the Treasury for each year each claim is held, with the amount of the fee increasing steeply after 5 years "so that idle speculators' claims will be cleared."

Once a mine operation is started, the MPC recommends a royalty be paid to the government that would amount to 12.5% of mineral production value, and no consideration would be allowed for profitability of the mine—in other words, a 12.5% tax on gross production value (which would be at least 25% of net income). In addition, miners would pay their normal federal, state and local taxes, and the cumulative total would put most of the currently profitable mining operations out of business.

Full reclamation of all mining disturbances would be required, with mandatory bonding set high enough to ensure "complete reclamation" prior to the initiation of any exploration or mining activity. The MPC believes that strict enforcement should ensure "that all mining is done under proper permit authority, that the permit conditions are observed and proper environmental monitoring is carried out, that royal payments are honestly computed and paid, and that final closure specifications are fulfilled and the site is restored to beneficial use." Given the already existing high mining costs, the environmental and regulatory requirements, and the present tax structure, then adding all of the costs and requirements associated with the MPC Mining Law proposal, all the enforcement provisions of this "law" become superfluous: *no mining could take place after existing profitable mineral deposits were mined out, so no enforcement would be required.* Of course, all small mining operations, placer dredging activities, and mineral exploration companies would cease to exist just as soon as such a law was enacted.

So, the "strengthening amendments" offered by the MPC for Sen. Bumpers' proposal, or the MPC Mining Law proposal itself, would both accomplish the same purpose: the elimination of mining in the United States.

In the *Conclusion* of the Mineral Policy Center mining law proposal, the MPC states:

> Will the Red, White 'n Blue Mining Law become reality soon? A year ago no one with a sense of history would have predicted success. But the winds of environmental reform are blowing strongly now, and remarkable things are possible. We will oppose superficial reform legislation—the 1872 Mining Law has no brain; it needs a full transplant, not cosmetic surgery. Soon. For the benefit of the nation's environment—and of its mining industry, too.

Translation: Abolish the 1872 Mining Law and eliminate those pesky miners for the benefit of the nation's environmental extremists. The only possible "benefit" for miners would be that there would no longer be any reason for having to put up with the constant attacks by radical environmentalists. If this analysis seems extreme, consider the following:

The MPC is advertising and distributing copies of ABC's *20/20* program entitled *The New Gold Rush* on videotape. This program was extremely biased against mining; it contained numerous misrepresentations and several outright lies. A recent Accuracy in Media (AIM) Report (Sept.-A 1989) titled "20/20 Turns Toxic" tore this program apart, pointing out the falsehoods, misrepresentations and blatant bias. The thrust of the program was that all mining is dirty, disgusting and dangerous, and that miners are pouring massive quantities of cyanide into the environment and people's drinking water. Yet the MPC is distributing this piece of trash as fact.

In the autumn 1989 issue of *Clementine,* MPC's Phil Hocker wrote an article entitled "Cyanide Spring—Heaps of Gold, Pools of Poison," in which he outlined the potential threat from cyanide use in mining and constantly re-emphasized that it was an "extremely deadly poison." Remember the Tylenol scare involving cyanide? Or the Jonestown massacre (mass suicide)? *Both were included in an article about cyanide use in mining.* The perceptions of the general public can be swayed easily: mention the buzzword—in this case, cyanide; rerun horror stories to remind everyone of same; then tie cyanide use to mining, and you produce a masterful reaction of hysteria in any uninformed person who might read the article. This type of deception produces a reaction similar to that produced by the *20/20* program: blind, unreasoning fear, followed by a public outcry for government action to eliminate these horrible, distasteful problems.

The MPC publishes articles praising those environmental activists who are successful in shutting down proposed mining operations before any mine development has taken place. Citing the "potential" for problems created by a mining proposal, they believe it is much better to eliminate the mining activity and thereby preclude the remote possibility that problems may occur. Calling these anti-mining radicals "unsung heroes," the MPC actively encourages them in their efforts to stop mining.

In its quest for more members the MPC makes this statement: "Hardrock mining generates twice as much solid waste annually as all other industries and cities in the nation." Did you know that the ordinary soil, gravel and rock covering this planet is "solid waste"? No mention is made of the fact that most of these materials moved by mining are common, ordinary, environmentally safe natural substances. The major implication here is that *everything* touched by mining activities becomes contaminated and is no longer usable by mankind. According to this line of reasoning, the entire natural, undisturbed surface of this planet is forever polluted and therefore unusable.

Now, all of this information comes directly from materials published in *Clementine, The Journal of Responsible Mineral Development,* by the MPC. In addition, the *Mining Conservation Directory '89* was published by the Mineral Policy Center in August 1989, and it is the MPC's guide to local action on mineral development threats. This directory is intended to provide access to local citizens' groups around America who are working to prevent environmental destruction from mining and oil development. In the autumn 1989 issue of *Clementine*, over 140 different organizations and offices are listed. In this manner, all anti-mining people throughout the country are able to contact and obtain the assistance of all other anti-mining people.

The MPC also maintains a large and growing library of information, slides and photographs on mining and oilfield impacts from sources throughout the nation. *All* of these items focus on the negative side of mining issues; *nothing* is available on the positive side.

This organization tries to maintain that they represent a movement toward "responsible mineral development" while doing everything in their power to facilitate the destruction of any and all mineral development. They are actually attempting to recruit miners to their cause by turning them against the very industry they work for.

Don't take the Mineral Policy Center lightly. They are meeting with and influencing our U.S. senators and representatives, they are testifying before congressional committees (when mining groups are excluded), they are actively supporting any and all anti-mining radicals in the country, they are providing damaging materials to the news media, and they are encouraging the use of civil litigation to halt mining activities.

Somehow, somewhere, sometime, all miners are going to have to become organized enough to effectively oppose the anti-mining movement that threatens to wipe out everything we've worked for. The very real dangers represented by the environmental mania that is sweeping the nation must be taken seriously. We can no longer afford to be complacent and uninvolved.

16
MPC Leads Effort to Abolish U.S. Mining Laws

THE MINERAL POLICY CENTER (MPC) is an environmentalist organization based in Washington, D.C., that was formed for the specific purpose of eliminating the 1872 Mining Law. Although they continue to say that their primary objective is to ensure responsible mineral development, *all* of their activity has been directed towards inflicting as much damage on the U.S. mining industry as they can. Towards this end, MPC personnel have provided erroneous and misleading testimony to the U.S. Congress in many committee hearings on mining law issues, and they actively lobby individual members of Congress regularly on the same issues.

In addition, the MPC publishes a quarterly magazine called *Clementine*, which is primarily distributed to members of Congress, environmental activist groups, state legislators and other politicians. Every edition of *Clementine* is devoted to attacking mining activities in the U.S. and pushing for "reform" of the 1872 Mining Law. The MPC "reform" of the mining laws includes a high federal royalty on all hardrock minerals (up to 12.5% of gross production value), a "rental fee" on all mining claims to be paid annually, mandatory bonding at high levels for all types of mining activities, restoration of mined lands to the natural state existing before any mining took place, high fines and penalties for any violations of their proposed new "laws," the ability to eliminate mining development wherever they don't want it to take place, and provisions for perpetual policing of mining operations by the general public.

At this time, their efforts to shut down mining in the U.S. are largely unopposed. No attempts are being made to expose the MPC for what the organization actually is: a mouthpiece for radical environmentalism. No efforts are being made to counteract the MPC's extremely negative portrayal of mining, or their misinformation, distortions and anti-mining bias being presented to the U.S. Congress and the American public.

Now, brace yourselves for the extremely damaging potential that these people represent. They are being taken seriously by the members of the U.S. Congress, many other politicians, and the general public. Representatives of the Mineral Policy Center are invited to attend *all* Congressional hearings on mining law issues and present their biased and erroneous testimony. MPC personnel regularly give speeches at environmentally oriented meetings, hearings, conferences, seminars, and in other locales. They are specifically invited to these events, and their anti-mining message is generally accepted as the truth. Very few individuals or organizations are willing to openly attack their anti-mining bias in a legislative or public forum.

And this is just the tip of the iceberg. The MPC also directly assists local community groups working on mining "problems"; they lobby federal and state agencies for "tougher controls" on mineral development, and they use the courts by filing lawsuits to halt all mining activity they possibly can. The MPC actively assists radical environmental groups in their efforts to eliminate all mining in the United States, and they support anti-mining actions taken by environmental extremists. Under the guise of what they purport is "responsible mineral development," they are doing everything in their power to shut mining down—permanently.

This year, they are poised for final victory in the elimination of the 1872 Mining Law. In the most recent issue of *Clementine* (Winter 1990-91), Philip M. Hocker (MPC president) states:

> The final days of the 101st Congress contained a flurry of actions related to 1872 Mining Law issues. Each of the proposals was narrowly defeated, but the overall effect was to raise Congressional awareness about the Law and create an excellent foundation for action in the 102nd (Congressional session).

He goes on to say that the votes necessary to pass mining law reform can be gotten this year and refers specifically to the new mining law bills to be introduced by Sen. Dale Bumpers and Rep. Nick Rahall.

Under a subheading entitled "Sham Reform," Hocker says:

> The Mineral Policy Center is particularly concerned that the vehemence of the opposition will lead some Congressmen to settle for sham reform, which will eliminate patenting and impose a modest royalty. This would sidestep the biggest problems with the law. True reform *must* also include: discretion to deny mining in sensitive lands, reclamation requirements and standards and bonding, public involvement, enforcement provisions, and a cleanup fund for abandoned hard-rock mines.

Even though Sen. Bumpers' mining law bill would have shut down mining in the U.S., the Mineral Policy Center testified in Congress last year that the bill didn't go far enough (for instance, the MPC said an 8% federal royalty was too low, and advocated an increase to 12.5%). Although Rep. Rahall's mining law bill would have had some major negative impacts on mining, the MPC objected to the absence of federal royalties, stringent penalties, strict enforcement and other harsh provisions being promoted by the MPC. So, it seems quite obvious that the Mineral Policy Center would not be satisfied with moderate mining law reforms, and they will be presenting testimony this year for police-state provisions in any mining law bill proposed in Congress.

In the meantime, the MPC continues to publish misleading and erroneous information in an attempt to gain further support from a misinformed Congress and a misled general public. For example, in the current edition of *Clementine* they state: "The miner pays no royalty for ores removed; about $4 billion worth of free hardrock minerals are taken annually from public lands."

This statement is both misleading and false. The $4 billion figure is gross production value of minerals produced, and does not take into account the cost of production—which would be at least half of the total value. So, hardrock minerals are not "free," and much of the production comes from private lands, so the total value is intentionally biased. In addition, *all* mines pay some form of state taxes on mineral production and *all* mines pay federal income tax. *Most* mines also pay royalties to property owners, joint venture partners or investors—sometimes all three—and these royalties are also taxed at the federal and state level (and, in some cases, at the county level). *All* mines also pay the property, sales and other taxes paid by everybody.

There are four more "Information Nuggets about the 1872 Mining Law" in the publication, and all of them are either misleading, taken out of context, delete relevant information, or contain false information. The item is obviously intentionally biased and intended to mislead the reader, yet it is presented as "fact."

The MPC concludes the 1872 Mining Law presentation by stating:

These provisions have been retained by an aggressive mining lobby, *abetted by thousands of petty hobbyists and speculators* [author's emphasis] throughout the West who see the 1872 Mining Law as a chance to get rich. The result is a constant threat to non-mineral values, an immediate drain on the Treasury, and a large and growing public liability for future cleanup costs. Mining generates vast amounts of groundwater pollution and hazardous waste. Reform of the 1872 Mining Law is long overdue.

What I want to know is this: (1) Where do they get the right to call all small miners "petty hobbyists and speculators"? (2) How can they flat-out lie by stating that there is "an immediate drain on the treasury"? (3) Just where is the "growing public liability for future cleanup costs" coming from, when U.S. mining is the most overregulated industry on the planet? and (4) Where is the MPC data proving that (in the present context) "mining generates vast amounts of groundwater pollution and hazardous waste"?

Unfortunately, American society has, for the most part, become so brainwashed by individuals and groups purporting to represent the public's concern on environmental issues that they are seldom questioned on their real objectives or the accuracy of their statements. As a result, any radical extremist who chooses to label himself or herself as an environmentalist is automatically accorded an inordinate amount of credibility and respect. They can literally lie before Congress and not be taken to task for it. They can make totally irrational statements and have them accepted as fact, without being required to provide one iota of proof.

Because of the perceived higher moral ground occupied by those who profess a genuine concern for the environment, most people seem to be afraid to contest their views—especially, politicians. It has come to the point where laws and regulations are being designed to address general perceptions and the individual preferences of a few vocal groups. Factual information is being ignored and critical decisions are being made on the basis of emotional appeal, regardless of whether it is right or wrong and in total disregard of the ultimate consequences.

In spite of this incredibly biased situation, miners have to get the truth to politicians, the media and the general public. We really have no choice. If there were ever a time when miners were threatened with extinction, that time is now.

17
Mineral Policy Center Launches Another Attack on Miners

THE MINERAL POLICY (POLICE?) CENTER, with the full support of Sen. Dale Bumpers, Rep. George Miller, Rep. Nick Rahall and Interior Secretary Bruce Babbitt, recently launched another all-out attack on the nation's miners by a series of news releases and the issuing of an MPC "report" on the purported dangers of abandoned mines to the general public and the country's environment.

According to the statement by Rep. Miller, chairman of the House Committee on Natural Resources, the Mineral Policy Center report, entitled *Burden of Gilt*, is going to serve as the basis for his efforts to permanently rid the U.S. of miners by "reforming" the mining law this fall. Miller said, in part:

> *Burden of Gilt* isn't going to suffer the fate of many reports issued in Washington. It isn't going to sit on the shelf: it is going to have a dramatic and immediate impact on public policy... *Burden of Gilt* puts the Congress, the Department of the Interior, and the American people on notice of a policy nightmare that has built up—literally, in many cases—for nearly a century and a quarter. We have a responsibility to address that environmental and public health menace, and the Committee on Natural Resources and the House will do so in the mining reform legislation we will write later this year...we need a comprehensive program, realistically funded by the industry that generated the waste and earned the profits.

Secretary of the Interior Bruce Babbitt said in his statement:

> The catastrophic consequences of abandoned hardrock mines, as documented by the Mineral Policy Center, provide a dramatic illustration of the urgent need for comprehensive reform of the 1872 Mining Law. This archaic law enables irresponsible companies to take the minerals and run, leaving a toxic legacy for taxpayers to clean up... Because the 1872 Mining Law does not require industry to pay anything for publicly owned minerals, there is very little money available to clean up abandoned hardrock mines...

Part of the press release by Rep. Nick Rahall states:

> The fact of the matter is that in the absence of adequate regulation, hardrock mining operations are often undertaken without due regard to the environment. The result is a proliferation of acidified streams,

abandoned mine waste dumps, open mine shafts and other hazards associated with hardrock mining and milling... Today, I join with Chairman George Miller to say that the will is there, at least on our part, to not only seek the reclamation of abandoned hardrock mines, but to insure that a new generation of these sites are not created by existing mining operations.

In the Mineral Policy Center news release accompanying the *Burden of Gilt* report, MPC President Phil Hocker said:

Mineral Policy Center knew the problem of hardrock abandoned mines was bad when we first commissioned this report. The problem turned out to be far worse than we imagined. We were shocked by the number of sites out there and by the profound environmental damage that has already occurred. The potential for further devastation of crucial water supplies is nothing short of frightening.

The MPC news release goes on to state:

There are 557,650 hardrock abandoned mine sites in the United States, *Burden of Gilt* reports, and cleanup will cost between $32 billion and $71 billion... Based on Bureau of Labor Statistics data, Mineral Policy Center calculates that for every one million dollars invested in reclamation of abandoned hardrock mines, 26 new jobs will be created. A fully operational, nationwide Hardrock Abandoned Mines Reclamation (HAMR) program with an annual budget of $400 million will create at least 10,000 jobs. Given that the entire hardrock mining industry currently employs only about 50,000 people, HAMR projects will represent a considerable boost to employment—particularly in areas where declining employment in mining and mining services has created large pools of experienced unemployed workers with skills that are directly transferrable to mined land restoration.

Establishing a comprehensive national HAMR program will help the communities where mining has occurred and the nation as a whole.

Well, now, isn't that something? Several members of the U.S. Congress and the Secretary of the Interior have fully accepted a report prepared under the exclusive direction of radical extremists who are known to be rabid opponents of mining and who only represent environmentalist groups that have an avowed mission of eliminating all mining activity in the U.S., and they are entirely willing to base congressional legislation and federal policy upon this thoroughly biased and misleading information. And, they are trying to pass it off as a measure to increase employment. It should be obvious that they fully intend to eliminate productive mining jobs and replace them with unproductive, state-run environmental jobs.

Even worse, guess how these people intend to pay for a nationwide abandoned mines reclamation program? The Executive Summary in the MPC *Burden of Gilt* report states, in part:

> While the mining companies have been reaping the vast fortunes that hardrock mining has long bestowed, the 'burden of gilt'—the degrading unreclaimed land abandoned by the mine operators—has been left to the public to deal with. Abandoned mines are silent killers, threatening public safety and health and creating long-lasting environmental hazards. Toxic mine wastes endanger people downwind, destroy aquatic life downstream, and contaminate vital groundwater resources. Abandoned mines constitute an intolerable threat to the nation's future...
>
> It is possible to design a pay-as-you-go HAMR program financed by various combinations of fees on minerals removed from private land and royalties and fees on minerals extracted from public land, augmented by penalties for violations of federal law. Examining a subset of the overall funding mechanism, such as a minerals royalty on federal lands, illustrates the funding potential of a HAMR program. For example, a 12.5% royalty (the same rate charged to producers of oil and gas on public land) coupled with an annual $100 rental fee for mining claims on public land should yield about $400 million annually...

Now, pay attention to this part in particular: The Mineral Policy Center is openly advocating that the abandoned mines reclamation program *be applied to both private and public lands,* and they want the federal government *to impose fees/royalties on minerals produced from private land, where the minerals are owned by private individuals.* This would place the nation's mineral-bearing private lands under total government control and regulation, place federal environmental and reclamation standards on previously mined property that is now in private ownership, and allow the government to impose fees and royalties on minerals that it doesn't even own. This part of the "report" alone adequately illustrates Phil Hocker's socialist agenda: state-controlled enterprises replacing the private business sector and superseding private property rights.

It also illustrates the government's and radical left's objective of imposing perpetual, unlimited liability on private individuals and businesses to address perceived problems (read: pay for) for which they have absolutely no responsibility. In addition, it again shows the tendency of big government and extremists to foist off the cost of desired programs onto the backs of selected individuals and enterprises (i.e., selective taxation for the benefit of others), (i.e., you pay for what I want).

Several qualifier statements are included in the report to address the HAMR funding mechanism, as follows:

> Whatever the fee schedule, revenue projections should be adjusted to reflect anticipated losses of production, if any, that might be precipitated by imposing additional costs on the industry..." (i.e., if you put most of the mining industry out of business by implementing this program, raise the fees on the remainder to ensure adequate revenues for the HAMR programs).
>
> ...However, given other variables that can have proportionately greater effect on profitability—recessions and recoveries, which drive consumer demand; trade agreements and dollar valuations, which affect minerals imports and exports; and multilateral bank lending policies, which affect Third World minerals producers' competitiveness—it is doubtful that a reclamation fee by itself would actually have much effect on most U.S. minerals producers...

What garbage! By making the statement above, Phil Hocker shows his total unmitigated ignorance concerning the conduct of any business enterprise, let alone the factors impacting the mining industry. He suggests the imposition of a 12.5% gross production royalty on minerals to pay for his so-called HAMR program, and then absolutely proves his stupidity by saying "it is doubtful" that this major economic hit would have "much effect" on the U.S. mining industry. And, don't forget Rep. Miller, Rep. Rahall, Sen. Bumpers and Interior Secretary Babbitt, who have swallowed this monstrosity hook, line and sinker.

Now, let's take a look at the heart of the gross assumptions made in the *Burden of Gilt* report: of the alleged 557,650 abandoned mine sites in the MPC report, 194,500 sites are listed as "benign" (typically need little, if any, further remediation); 231,900 sites are listed as "landscape disturbance" (need landscaping/revegetation to prevent offsite "effects"); and 116,300 sites are listed as "safety hazards" (need prompt, but not necessarily extensive, action). Quick summary: benign, 34.8% of total sites; landscape disturbance, 41.5% of total sites; and safety hazard, 20.8% of total sites—or 97.1% of the total 557,650 sites estimated. Just where in the hell is the national calamity outlined in the news releases and the report?

A total of 14,950 sites, or 2.7% of the total sites, are listed as having any major environmental consequences (surface water contamination, groundwater contamination, and Superfund sites [50]). So, only about 2.7% of the sites require actual remediation beyond public safety factors (normally solved by adequate fencing and posting of hazard warnings to the public). Many of these problem sites are already being worked on. Of the first three categories (97.1%), a large number of sites are already being reclaimed and/or made safe by the mining industry, individual miners, and state and county agencies. *No mention of these facts is included in this report.*

For example, the Nevada Abandoned Mine Lands (AML) program to address physical hazards to persons or animals has been operating since 1987, has estimated 50,000 hazardous conditions in Nevada, and has inspected and documented 5,458 mine sites, and 2,780 sites have been secured by either mine owners, the Department of Minerals, Good Samaritans, or county agencies *at an average cost of $425 per site* (the MPC and Phil Hocker estimate a cost of *$19,500 per site* in this category). In addition, only 11 (0.2%) of the documented sites showed potential for chemical toxicity, and only 52 (1%) of the documented sites contained any water at all. The Nevada AML program *is funded entirely by the mining industry and has an annual budget of $250,000, and it has been functioning very well for 6 years—without any assistance from the Federal Government, environmentalists or Phil Hocker.* In addition, Nevada's miners are already performing their own reclamation and remediation as well as voluntarily reclaiming nearby abandoned mine sites for which they were not responsible.

By the way, under Acknowledgments in the *Burden of Gilt* report is listed the following: "Philip M. Hocker, President of Mineral Policy Center, whose deep commitment to responsible land use and firm sense of editorial direction *guided this report from conception to completion...*"

Need we say more?

18
Mother Nature Charged with Environmental Crimes

Author's Note: This article is a *fictional account* of what would happen if environmental extremists and government agencies attempted to subject Mother Nature's natural actions to the same environmental laws and regulations that are applied to human activities. In other words, it is a general comparison between the real world and natural events versus actions and impacts of human beings living and functioning on the planet. While the natural events described herein are factual, their net and cumulative impacts have, necessarily, been estimated.

Editor's Note: The years in which the fictional events in this satirical article occur have been changed (pushed further into the future than the author designated them when he wrote the article in 1993) to more accurately reflect the futuristic thrust of the piece.

IN THE YEAR 2025, environmental groups and government regulatory agencies had finally achieved their ultimate objectives in the United States. They had effectively eliminated most of the pollution and environmental degradation caused by business, industry and other productive human activity. The so-called "rape and ruin" of America's land and water resources had been brought to a screeching halt by doing away with all of the enterprises that might have a negative impact on the environment.

The "Vision of Change for America" program had, by now, been fully implemented: There were no more ranches, mines, sawmills, foundries, chemical plants, smelters, refineries, coal- or oil-fired power plants, or similar facilities, and the agricultural sector had been fully converted to organic growth methods (without pesticides or fertilizers). All of the manufactured goods, finished materials and necessary maintenance supplies were now being imported from foreign countries. Most of the nation had become one vast national park, and the Federal Environmental Protection Agency (EPA), which was now a separate branch of government employing 25% of the total national workforce with an annual budget of $4 trillion, was engaged in a massive effort to restore America's lands and waters to a "pristine state."

Poised on the verge of fully realizing their utopian ideals, the nation's environmental groups were slowly beginning to discover that the pollution and degradation of the environment had not yet been stopped. Even in those areas where all human activities had been eliminated (about 70% of the U.S., including all of the Pacific Northwest, Alaska and the Rocky Mountain states), the EPA standards for air and water quality still could not be met. And, horror of horrors, the holes in the ozone layer still appeared with alarming regularity. To top it off, many of the 951,813 threatened and endangered species listed were still inexorably being wiped out, despite the U.S. Fish and Wildlife Service's expenditures of $100 billion annually to protect them.

Environmental leaders quickly mobilized their members and, after informing them of the urgent need to mitigate the problems, sent them into the field to locate the culprits responsible for these crimes against nature. After several months of frantic searching, however, they were unable to pin the criminal actions on anyone. Worse yet, they had discovered mounting evidence that a far larger number of violations of environmental laws and regulations were being committed.

Frustrated in their extensive efforts, the environmental leadership immediately arranged a series of meetings with EPA officials and specialists and presented them with the now greatly expanded list of criminal violations. The EPA launched a massive investigation into the matter and after two years of study at a cost of $78 billion, finally issued an official report: these atrocities were entirely due to natural causes! Ultimately, Mother Nature was held to be responsible.

The Natural Resource Defense Council immediately filed a civil action against Mother Nature in the courts, asking for $10 billion in direct compensation and $700 billion in punitive damages. At the same time, the EPA filed criminal charges against Mother Nature in federal courts, specifying $18 trillion in fines, demanding an immediate cease-and-desist order against all natural events found to be in violation of environmental laws, and recommending the maximum prison sentence (approximately two million years) as well as the maximum in punitive damages (roughly $4 quintillion).

Shortly thereafter, the U.S. Congress passed the omnibus "Mother Nature Crime Act of 2028" (H.R. 8990031579244), in which 1.5 million specific known criminal violations of environmental laws and regulations were enumerated. Several aspiring politicians also began to run election campaigns with a platform based upon "doing away with Mother Nature." Jumping on the bandwagon, Congress quickly enacted 4,312 new environmental laws—just to make sure that all the bases were covered.

In order to fully investigate this important matter, a National Board of Scientific Inquiry into Natural Environmental Degradation (NBSINED) was quickly authorized. The board's membership consisted of 1,012 environmental radicals and 3 environmental scientists. Funding for NBSINED's critical work was set at $2 trillion annually. Because the federal income tax rate already stood at 75% of gross income and federal revenues had dropped by roughly 96% (with a current total national debt of $878 trillion and a foreign debt of $775 trillion), the government continued to print massive amounts of unbacked paper currency to pay for these new programs (however, by this time most notes were in denominations of $1-, $10- and $100-billion, with a monthly inflation rate of about 5,051%).

After 5 years of intensive study, in late 2033 the NBSINED issued a preliminary report, entitled "4 Billion Years of Environmental Degradation: Mother Nature's Crimes against Nature," which was 21,402 pages in length. Some of the findings and conclusions included in the report are as follows:

1. *Species Extinction:* Considering only the known data (additional research was continuing), the report found that Mother Nature had maliciously and deliberately caused the extinction of well over 3.5 million species of plant and animal life throughout the history of the planet. Under the Endangered Species Act (ESA), which had been back-dated to cover the time period encompassed by the report, this would constitute an uncountable number of "takings" of individual members of each "threatened and endangered species" which, by definition, would include all of the extinctions to date. Even more important, the report found that numerous species were currently undergoing natural extinction under Mother Nature's "survival of the fittest" program. For that matter, it was discovered that quite a few protected species were actively contributing to the extinction of several other threatened and endangered species through natural predation.

Conclusion: An additional $400 billion annually should be immediately appropriated for species protection and designation of vast new regions as "critical habitat."

2. *Volcanoes and Volcanic Vents:* The report listed evidence which showed that Mother Nature had been indiscriminately creating literally millions of volcanoes and volcanic vents throughout geologic time, and she was still engaged in the practice of creating even more of same. Not only were many of these volcanic structures directly responsible for the extinction of various species of plant and animal life, every single one of them was in violation of EPA emission standards for noxious gases, toxic heavy metals and airborne particulate materials. Further, it was found that the type and tremendous volume of venting volcanic gases was contributing directly to the depletion of the ozone layer. In addition, it was discovered that a number of violent eruptions had caused changes in the global climate and might have, during periods of intense volcanic activity, actively assisted in the onset of several Ice Ages. For some strange reason, the tremendous destruction caused by lava flows, ash falls, and volcanic "bombs" was not even mentioned in the report. However, studies are ongoing.

It was reported that currently active volcanoes and undersea vents were annually expelling over 5 times the toxic emissions produced over the entire history of the human Industrial Age, and of over 100 trillion times permissible levels each month. The EPA maximum allowable emission standards, measured in parts per million (ppm) or parts per billion (ppb), were virtually meaningless: carbon dioxide, carbon monoxide, sulfur dioxide, nitrous oxide and other noxious gases as well as arsenic, antimony, lead, mercury, selenium and numerous other toxic metals were being ejected by the *ton!* (The overwhelming nature of this problem soon became obvious.)

Conclusion: A scientific investigation (using real scientists) to study the feasibility of blocking this massive pollution, funded at $200 billion minimum annually, should be started immediately, and at least $50 trillion should be appropriated to "cap and wrap" these environmental monsters if it was at all possible to do so.

3. *Disintegration and Erosion of Natural Materials:* The report estimated that at least 5 billion tons of natural pollutants (rock, sand, silt, clay, gravel and organic matter) were being unceremoniously dumped into North America's rivers, streams and lakes each year—and Mother Nature had not even applied for an EPA NPDES permit!

An analysis of the pollutant source materials revealed that she was disintegrating rock and organic materials at the rate of approximately 200 billion tons per year on a worldwide basis. Absolutely no attempt was being made to stabilize these processed materials or retain them in tailings or settling ponds, so the pollutants were being spread virtually everywhere throughout the natural environment.

The bulk of the pollutants were eventually being eroded into rivers and streams, and thence into the world's oceans. This illegal disposal of such a tremendous volume of natural waste materials was mind-boggling: How could the EPA ever hope to clean up such massive environmental degradation? Worse yet, how could the agency devise new regulations to prevent additional pollution from these sources?

It was noted that these disintegrated natural materials often became airborne during windstorms (particularly in dry climates), and the quantity of resultant suspended particulates quite frequently were in gross violation of the Clean Air Act (a computer simulation produced a rough estimate of 1.5093 trillion violations annually in the United States alone). When considered in the context of decades of such activity (let alone millions or billions of years), the cumulative air pollution became astronomical—thereby dwarfing the amount of airborne particulate matter produced by humankind over their entire existence on the planet (by a factor of near infinity).

Worse yet, virtually all of this particulate matter eventually ended up in water, thereby producing an overwhelming number of violations of the Clean Water Act. A 2-year study of Alaskan rivers and streams alone showed well over 750 billion separate violations of the allowable quantity of suspended particulate matter (both organic and inorganic) in the state's waters per year. Over time, in Alaska alone, the number of EPA water quality violations also approached infinity. A computer analysis of the waterborne suspended particulates produced by Alaskan miners throughout the entire history of placer mining in the state showed they were responsible for roughly 0.000000000001% of the total water pollution from this source (fortunately, all Alaska placer mines had been put out of business by the EPA 143 years earlier, in 2020).

Conclusion: That a $50 billion scientific study be authorized to investigate various means of "cementing" continental materials, in order to prevent their disintegration and erosion while still allowing plants to grow. (A rumor recently surfaced that a member of the NBSINED asked this question: "How can naturally occurring silt, sand, clay, and gravel be classified as 'pollutants?'" An EPA official answered, "It's in the regulations." This ended the matter, and the question never came up again.)

4. *Major Storms, Lightning and Precipitation:* The NBSINED report determined that Mother Nature's hurricanes, typhoons, tornadoes, cyclones, thunderstorms and windstorms were causing extensive damage to the ecology of the planet, in addition to being well out of compliance with numerous environmental laws, rules and regulations. Many major storms were also responsible for "takings" of threatened and endangered species, and some of them wiped out entire ecosystems as well as critical habitats for protected species. (Intense lightning storms have a nasty habit of turning cute and cuddly creatures into crispy critters.)

It was noted that heavy precipitation almost always produced criminal violation of the Clean Water Act, often degrading water quality to the point where waterways carried nothing but a muddy, sludgy soup. This was a direct result of accelerated erosion of naturally occurring pollutant materials. A determination was made that, overall, heavy storms were effectively outlawed by existing environmental laws, so they should not be allowed to continue to occur anyway.

However, on the flip side, it was also found that an almost total lack of precipitation (drought) was in blatant violation of the "no net loss" policy for the "wetlands." Drought also set up conditions for major air quality violations during windstorms.

Lightning storms were found to be responsible for causing thousands of forest, brush and grass fires each year, sometimes burning down huge areas of protected "old growth" forest and creating unacceptable damage to numerous sites within the National Wilderness Preservation System. In addition, indiscriminate thunderstorm activity was producing massive quantities of ozone pollution in direct violation of EPA standards (ozone in the lower atmosphere is a "pollutant," while in the upper atmosphere as part of the ozone layer it is "good ozone"). It was also determined that the EPA standards for carbon monoxide and carbon dioxide emissions, as well as EPA standards for airborne particulate matter, were being exceeded by tremendous amounts during all lightning-caused fires. Some of these fires were creating more pollution than that caused by business and industry before the EPA shut them down in 2023.

A computer simulation of Mother Nature's atmospheric activities, and the resultant impact on the planet's ecology, lands, and waters, [indicated that they] were responsible for roughly 8 quintillion major violations of environmental laws and regulations every year!

Conclusion: An appropriation of $5 trillion should be made to fund scientific research on "weather control" techniques, with the desired objective of eliminating major storms of all types and developing technology to produce slow, even and limited precipitation wherever and whenever it was needed to enhance the planet's ecosystems.

Other important sections of the NBSINED report dealt with Mother Nature's civil and criminal violations of environmental laws and regulations in the following areas: (a) impacts of meteorites, comets and asteroids on the Earth throughout 4.5 billion years of operations, and their resultant effects on the environment; (b) tectonic plate movement, earthquakes and related volcanism and tsunamis, and their environmental impacts over time; (c) the emplacement, distribution and atomic disintegration (decay) of radioactive elements, with special emphasis on volatile isotopes such as radon, and the resultant effects on Earth's plant and animal life forms, and (d) solar activities, such as sunspots and variations in heat and the emission of radiation, and their overall impacts on the planet over time.

Following the release of the NBSINED report in early 2034, the horrendous extent of Mother Nature's environmental crimes became a top government priority. Over 100,000 lawyers were employed full-time in efforts to file civil, criminal and financial penalties against Mother Nature for each and every one of her violations of environmental laws and regulations. In the meantime, the President of the United States declared a national emergency and deployed the entire U.S. Armed Forces (numbering exactly 12,102 personnel by this time) to address these problems. The U.S. Attorney General immediately recommended limited paroles for the 25% of America's population that had been imprisoned for environmental crimes so they could be utilized in massive "cleanup" projects. Although the EPA was, at first, vehemently opposed to the release of "environmental criminals" under any circumstances (most of them had received life sentences), they eventually relented under the condition that the parolees be organized into "prison gangs" under the immediate supervision of "eco-police." The nation's environmental groups immediately volunteered to be guards.

As usual, Mother Nature continued to conduct uninterrupted operations despite the civil and criminal actions being lodged against her—as she has been doing since the universe was formed.

19
Anti-Mining Propaganda: Half-truths and Lies

CMJ **Editor's Note:** One of the most hard-to-swallow developments for mining in the 1990s has been the media bashing of the industry. It is not only a threat to the livelihood of those that derive their living from mining and ancillary products and services, both large and small, but it insults one's sense of honesty and fair play and has placed many of us on the defensive, even with our neighbors.

As we usher out the old year and welcome in the new, we thought it appropriate to dedicate more space than usually allocated to the "Bawl Mill" to address this issue. The following was authored by Dave W. Parkhurst.

THE TOTALLY UNCALLED-FOR trashing of America's minerals industries and miners by the media and anti-mining radicals has again become fashionable, as evidenced by the recent flurry of biased and untruthful articles about mining that have appeared in normally credible national publications as well as the intentionally inaccurate anti-mining programs aired on the national television networks.

These negative portrayals of mining are intended to perpetuate the myth that miners are engaged in the controlled destruction of our nation. Nothing could be further from the truth. Both federal and state agencies enforce a multitude of strict mining regulations and laws that are more than adequate to protect our country's environment. In fact, mining and minerals-related activities are among the most overregulated enterprises in the United States.

The picture presented by these trashings of a legitimate and productive business is also intended to lend support to the general public's misperception that mining is dark, dirty, distasteful, dangerous and destructive, as well as the fantasy that miners are intentionally engaged in the rape and ruin of our lands. A more inaccurate and untrue depiction of mining and miners is difficult to visualize.

What is the ultimate objective of this media blitz, and who is orchestrating the hate-mining campaign? The rapid destruction and elimination of all mining activity in North America is the objective, and the anti-mining propaganda campaign is being orchestrated and funded by radical extremists in the so-called "environmental movement." Because they realize they cannot achieve their goal by presenting the truth to the American public, these extremists have resorted to the dissemination of propaganda based upon biased reporting, half-truths and outright lies. They believe (and they are correct) that the repetition of lies in the media, and the lack of any truthful response, will eventually get the American public to accept their propaganda as the truth.

It might be helpful to provide a definition at this point: *Propaganda:* (1) a systematic effort to persuade a body of people to support or adopt a particular opinion, attitude, or course of action; (2) any selection of facts, ideas, or allegations forming the basis of such an effort; and (3) an institution or scheme for propagating a doctrine or system. *Propaganda is now often used in a disparaging sense, as a body of distortions and half-truths calculated to bias one's judgment or opinions.*

For example, Adolph Hitler, Joseph Stalin and several other dictators utilized propaganda quite successfully in their efforts to brainwash the general public, so the method has an historic record in its use and application.

Some other people have noted the similarities between the recent media blitz and organized propaganda campaigns. For instance, Paul S. Strobel, editor of *Mining World News,* notes in an editorial in the November/December issue:

> One can't help but suspect motivations and political manipulations when reading rhetoric so familiar as that presented in this latest assault on the (mining) industry.
>
> Mining has come to expect unfavorable dialogue from preservationist newsletters and environmental journals. But the past year has seen presentations from some really heavyweights in the media: for example, e.g., *20/20, Readers Digest, Newsweek,* and now *U.S. News and World Report*.... The media coverage is filled with the same half-truths and slanted views expounded by certain elite circles of mining opponents one hears from in special interest newsletters and protest rallies.

Strobel says it is obvious that one or more of the following situations are occurring: (1) the media is being unknowingly manipulated by special interest opponents of mining (he finds this hard to believe); (2) the media is so anxious for readership, or viewership, that objectivity is sacrificed for the sake of front-page sensationalism and provocative headlines (he thinks this is quite likely and that it is certainly understood by promoters of anti-mining causes), and (3) the media is knowingly promoting controversy in concert with certain mining opponents for both profit and in a misguided attempt to bring about change in an industry they neither understand nor wish to objectively research (this is also likely).

Strobel believes that no matter what the reasons are, the general public is getting *"a slanted, misinformed view of mining—and this impacts congressional representatives, regulators, and administrators at all levels of government."*

These concerted attacks on mining by the major national media, although a relatively new phenomenon in normally credible publications, are far from being isolated incidents in the news media. A number of major newspapers have been consistently taking below-the-belt shots at mining for many years by biased reporting and the publication of deliberately misleading information, half-truths and certain sensationalistic lies. Unfortunately, this illustrates a somewhat pervasive anti-mining mindset in many of the people who control, and write, the information published (or aired) by the media. Certain selected businesses and industries have been subjected to constant harassment (and trashing) by the news media for about the past three decades, and the constant repetition of this propaganda has slowly, but surely, brainwashed the American public into believing most of this garbage. This has already produced some severe negative impacts on the nation's business and industry—particularly mining.

The most frightening aspect of this situation is the major media's absolute refusal (in most cases) to publish the truth, *even when confronted with irrefutable proof that they are publishing untrue information.* This means that a large number of editors and reporters are deliberately choosing to mislead and misinform the American people. It would appear that the latter definition of propaganda listed above would be an appropriate description of the recent media attacks on mining.

One of the first casualties suffered by the industry in this "war" on North American mining has been the effective elimination of mining's ability to plan for the future—not just in terms of years but also in terms as short as weeks or months. Mining is a business just like any other, and there must be some basis for planning business operations. Mining can no longer do this, because current legislative and regulatory actions might make it impossible for the industry to function at any time. No one in their right mind is going to expend money on a project that has a risk of being shut down, permanently, at any time.

In the November 1991 Northwest Mining Association Bulletin, NWMA President David A. Holmes said, in part:

> But government actions based on pure political reasons are the one area that we cannot predict, plan for or understand. *The result has been a partial collapse of a vital industry and withdrawal of financial support.* The contest between good geology and bad government is a tilted playing field, in favor of the unknown.
>
> The real loss is the nation's basic ability to discover and produce minerals domestically, for the economy, security, jobs and community survival. Political decisions are safe for at least two reasons: the political fall-out seldom affects the authors of the decision, as they enjoy continuing paychecks regardless of its quality or sense; and secondly, those decisions are too often pushed by a political agenda funded by a part of society that has been led to believe that mineral production happens in the surroundings of 1891 rather than 1991. Even though the pendulum really is beginning to return to the center, the pace is painfully slow.
>
> Our 'Job One' is to help the pendulum return to the middle. People are fed up with lawmakers and their staffs for not taking...responsibility for their actions. *People are sick of the excesses of anti-business groups that can continue publishing lies about mining and the mining laws.*

A few of the most outrageous media statements are listed below. They are hereby nominated for a new award, called the "Award for Best Media Lie/Distortion Published in 1991."

1. "[L]oopholes (in the 1872 Mining Law) have spawned a 'Great Terrain Robbery,' turning public resources into a haven for land speculation and environmental degradation" (from "The Great Terrain Robbery," by Randy Fitzgerald in the April 21, 1991 issue of *Reader's Digest*).

Fact: According to the Department of the Interior, less than one quarter of one percent of 662,000 active mining claims in Arizona, California and Nevada have known or suspected unauthorized activities, almost all of which are by nonminers. This is better than the average crime rate in most large cities every year and, over time, there is no comparison.

2. "But miners of gold and other hard minerals operate under an 1872 law as antiquated as a prospector's pan. Anyone can file a claim and mine it for next to nothing... You don't need to strike gold to get rich. Claimants who invest at least $100 a year can buy sites for as

little as $2.50 an acre and resell them..." (from "The War for the West," by Bill Turque in the September 30, 1991 issue of *Newsweek*).

Fact: The 1872 Mining law is not antiquated, it's been amended and/or affected by at least 37 legislative actions; and operations under the law are directly regulated by numerous federal, state and local regulations. (The prospector's pan isn't antiquated, either—it is still as useful as it was several centuries ago.) Recent estimates to patent (purchase) mining claims average $12,000 per acre by the time the process is finally completed. Some of this is for lands with a full market value of around $50 per acre.

3. "To fly across areas like north-central Nevada is to view an alien landscape carved by mechanical forces on a Brobdingnagian scale... But there are no requirements for environmental protection and reclamation... 'This is destruction on a truly heroic scale,' says Sierra Club mining expert Glenn Miller..." (from "The New Gold Rush" article by Michael Satchell in the October 28, 1991 issue of *U.S. News & World Report*).

Fact: Mr. Satchell must have X-ray vision: less than two-tenths of one percent of the lands in this area has been disturbed by mining activity. A multitude of environmental and reclamation laws and regulations at the federal, state and local levels apply directly to all types of mining activities, and they are strictly enforced. "Destruction on a truly heroic scale"? Come on, Glenn! In addition, Glenn C. Miller is a biochemist at the University of Nevada, Reno, and *he has absolutely no mining experience in any way, shape or form*. Mr. Miller is certainly no mining expert, but he *is* a renowned Sierra Club activist.

More Facts: All mining activity in the United States for the past 200 years or so has only impacted about 26 one-hundredths of one percent (or .0026) of the nation's total land area (includes coal mining and gravel pits). About one-third of that land disturbance has already been reclaimed, and more is being reclaimed at a rapid rate. In contrast, America's towns and cities cover nearly 35 million acres (or 1.5% of the total land area)—well over five (5) times the total area mined. And mining provided all of the materials necessary to build those towns and cities. In addition, the nation's system of roads and highways covers roughly 24 million acres (or 1.05% of the total land area)—over four (4) times the land area disturbed by mining, and concrete and asphalt paving creates a much more permanent impact on the environment than does mining.

Space limitations preclude complete coverage of this matter; at least a book-length manuscript would be required. It is hoped, however, that sufficient information has been provided herein to accurately portray the media problem. Miners *must* take action to counteract these media attacks on mining or the anti-mining propaganda campaign will achieve its objective of putting us out of business.

20
Miners Show Strength and Unity at Mining Law Hearing

MINERS SHOWED STRENGTH and unity during a sometimes testy, daylong hearing held September 6, 1990, before the House Mining and Natural Resources Subcommittee. The subcommittee was considering legislation that would completely replace the 1872 Mining Law. Chairman Nick Rahall (D-WV) had introduced H.R. 3866, the "Mineral Exploration and Development Act of 1990," on January 23, and this was the first hearing held on the mining law proposal.

Twenty-six representatives of mining groups, associations, companies and individuals testified that the 1872 Mining Law, as amended, has continued to work well for over 118 years, and that any complete change, revision or replacement of the law is both unnecessary and counterproductive. The miners' testimony was presented in a clear, factual and comprehensive manner, with each individual's oral and written testimony fully supporting the testimony provided by other witnesses on behalf of the nation's mining community.

In contrast, nine representatives of environmental organizations depicted mining and miners as being insensitive to environmental concerns and almost totally unregulated. Their testimony concentrated on the need for extremely strict and onerous laws and regulations, recommending that many of the unrealistic and impossible demands by anti-mining activists be included in any legislation enacted to revise U.S. mining law. They also advocated changing to a mineral leasing system with high federal royalties on mineral production.

For example, consumer advocate Ralph Nader and Mineral Policy Center president Phil Hocker recommended the addition of Sen. Dale Bumpers' onerous provisions in S. 1126 and even more stringent proposals from the Mineral Policy Center, including high federal royalties, a minerals leasing system, and public discretion on whether to allow mining to take place at all.

Miners countered with testimony that showed mining to be environmentally responsible and listed the overwhelming number of laws and regulations that apply to all types of mining activity today. In addition, the mining representatives presented factual analyses of the potential negative impacts of H.R. 3866 if the bill were to be enacted, showed how well minerals exploration and development is taking place under current law, and pointed out numerous errors and misrepresentations in the testimony given by environmental groups.

They were ably assisted in their efforts to present an accurate picture of mining in the U.S. by Rep. Larry Craig (R-ID) and Rep. Barbara Vucanovich (R-NV). Rep. Craig pointed out that the mining law has been changed 55 times since it was originally enacted and that it is therefore not antiquated and needs no extensive revision or replacement. He also questioned environmentalists on their testimony and emphatically pointed out several errors and misrepresentations.

Rep. Vucanovich stated her total opposition to changing the mining law and introduced a letter in support of the current mining law that was signed by all members of Nevada's Congressional delegation. She also introduced the petitions signed by over 17,000 miners (CMJ petitions) opposing changes in the mining law.

BLM Director Cy Jamison testified that revision of the 1872 Mining Law, as amended, was not required and said the BLM could handle any necessary administrative actions by regulations and proper enforcement of existing laws. However, he did note that the BLM will be requiring mandatory bonding for land disturbances of less than 5 acres in the near future. He further noted that the agency will also clarify the definitions of uncommon varieties of minerals, tighten regulations on the occupancy of mining claims, and consider raising the fee per acre for mining patents.

A Nevada mining representative also introduced into the record a personal letter from Nevada Governor Bob Miller stating his full support for the 1872 Mining Law and total opposition to enactment of Rahall's H.R. 3866 as well as his active support of the testimony provided by an individual representing Nevada's mining interests.

Rep. Nick Rahall, as subcommittee chairman and author of H.R. 3866, maintained that at least some changes to the 1872 Mining Law are necessary. However, he questioned both miners and environmentalists carefully on their testimony and asked them for their opinions on various provisions of his proposed legislation. He noted that no further action will be taken on the bill this year but said he would revise the proposal and reintroduce it in the next session of Congress.

At the close of the hearing, mining representatives felt they had a fair hearing, and they were generally positive in their assessment of the probable outcome of their efforts. Almost all of the miners present said they believed the primary objectives of the group—namely, to present a comprehensive and accurate picture of mining to Congress and to present an effective opposition to the revision or replacement of the 1872 Mining Law—were accomplished in the best possible manner.

21
Miners Dominate Reno Mining Law Hearing

THE SUBCOMMITTEE ON MINING and Natural Resources of the House Committee on Interior and Insular Affairs held a field hearing on proposed reform of the U.S. mining laws in Reno, Nevada, on April 13, 1991. The subcommittee chairman, Rep. Nick J. Rahall (D-WV), scheduled the hearing to hear testimony on H.R. 918, the "Mineral Exploration and Development Act of 1991," which he had introduced in the U.S. House of Representatives last February.

Only those witnesses invited by the subcommittee were allowed to present oral testimony, but all other interested parties were allowed to submit written testimony. Those invited to testify included state and local government officials, small-scale miners, mining companies and associations, environmental groups and anti-mining activists. Of the 45 scheduled witnesses, 30 testified in favor of mining (21 miners and 9 state and local officials) and 15 were in opposition (mostly environmentalists and a few anti-mining groups). The audience was pro-mining, as were several hundred miners who attended a rally outside the hearing room organized by People for the West! in support of mining.

The hearing testimony was also overwhelmingly pro-mining, with state and local officials providing solid opposition to H.R. 918 and full support for Nevada's mining community. Nevada Governor Bob Miller was the lead-off speaker, and he set the tone for the balance of the hearing. His opening remarks emphasized the importance of mining to the state of Nevada and included statistics to illustrate the contribution Nevada's miners make to both the state and the nation. He then addressed the mining law issue, stating:

> A healthy mineral industry is not only important to our state, but to the nation as well. Reliance on imported mineral materials continues to increase at an alarming rate. The security of our country is weakened whenever domestic production of a commodity is lost to a foreign supplier.
>
> We are opposed to the bill under consideration today, H.R.918, because it would damage the ability of the mining industry to develop minerals from public lands. The bill goes too far in its attempt to address what have been perceived as problems with the General Mining

Law, as amended. Russ Fields, Executive Director of the State Department of Minerals, is here today and he will identify specific issues in the bill that present problems.

The State of Nevada's position is that the General Mining Law, as amended, should continue to be the foundation of mining law in this nation. The concepts of self-initiation, free access and security of tenure have stood the test of time and have served well. There is no reason to change those concepts. The federal and state agencies have the tools they need to effectively regulate exploration and mining.

We believe that areas of the mining law which are of concern to members of Congress should be more thoroughly analyzed and, if changes are required, those should be addressed through amendments to the existing laws, regulations and policies. However, any amendments should only be undertaken after careful consideration is given to the potential impacts those changes would have on the long-term health of the nation's mining industry.

This matter is extremely important to many of our Western States. So important, in fact, that I recommend formal lines of communication be opened up between the West and your office, Mr. Chairman, and Senator Bumpers' office. We have the expertise to work with the appropriate Congressional staffs to analyze the concerns you have and to recommend effective approaches to any required changes.

I ask that the Subcommittee give strong consideration to exploring more effective ways to address the concerns expressed by the Chairman and others about the Mining Law. I am certain that I and other Western Governors can recommend representatives from the West to work with the appropriate Congressional staffs to identify real problems and posit effective solutions.

Immediately following the governor's presentation, miners and other supporters gave him a standing ovation. Of particular note was the fact that individual anti-mining and environmental activists remained seated during the ovation, thereby positively identifying almost all of the anti-mining faction present at the hearing.

Panel I followed the governor, and it consisted of: Larry Beck, Washoe County commissioner; Sammye Ugalde, Humboldt County commissioner; and Russ Fields, executive director of the Nevada Department of Minerals. All three officials firmly opposed H.R. 918 and fully supported the state's mining industry. Fields presented a section-by-section analysis of the problems contained in the legislation, as well as a comprehensive overview of the total projected impacts of the bill if it is enacted.

Panel II consisted of seven small-scale miners who represented themselves as well as three small-scale miners' associations. Every member of the panel provided accurate and effective testimony, and several individuals received ovations from the audience following their testimony. Hugh C. Ingle, Jr., a small mine operator and first president of the Nevada Miners and Prospectors Association, was particularly effective in his testimony, and he received vocal support from the audience on several occasions during his presentation as well as an ovation when he finished. Others on the panel were John Livermore, Merle Swanson, Dave W. Parkhurst, Don Smith, Mack Taylor and Patricia A. Holmberg of the Idaho Mining Association.

Panel III consisted of five representatives of environmental groups and activist groups, most of whom concentrated on the resurrection of mining problems from the distant past. Very few actually addressed specific provisions in H.R. 918, except to say that the environmental and reclamation requirements were not nearly as stringent as they would like.

The six panels that followed (four pro-mining and two anti-mining) basically followed the pattern set by the preceding three panels and the governor. Environmentalists continued to say that the 1872 Mining Law allows major environmental damage, as well as the destruction of wildlife and wildlife habitat, and that it produces maimed landscapes and rampant pollution. Miners and mining industry representatives continued to point out the potentially disastrous contents of H.R. 918 and the fact that it would signal the permanent decline of mining in the United States.

Throughout the hearing, Rep. Rahall insisted that mining law "reform" was necessary and that he would continue to promote legislation to accomplish that objective, but he repeatedly stated that the provisions in the bill are not set in stone and he is willing to make necessary changes. The only two subcommittee members attending the hearing were Rep. Rahall and Rep. Barbara Vucanovich, who is a stout defender of miners and existing U.S. Mining Law.

Rep. Vucanovich maintained that the 1872 Mining Law works well and does not need any "fixing." She received several standing ovations from the audience for her unyielding defense of the mining law. Vucanovich also consistently rebuffed anti-mining statements by environmentalists and emphasized the fact that the 1872 Mining Law is a land tenure statute, not an environmental statute, and that all hardrock mining activity is subject to a multitude of existing environmental laws and regulations.

At the end of the hearing, the miners who attended as participants and observers generally agreed that those who testified had done the best job possible in their attempt to address the legislation. However, it was also agreed upon that the effort was probably not sufficient to defeat the bill or make meaningful changes in its content at the subcommittee level. As a result, they are going to increase their efforts to obtain support from each individual member of Congress and other high-ranking politicians to kill the measure if it comes up before a full vote in the U.S. House of Representatives and the U.S. Senate.

There was another interesting event that relates directly to the hearing objectives set by Nevada's small-scale miners. On the day following the Reno hearing, April 14, Rep. Rahall and Rep. Vucanovich, with staff, were taken on a tour of Hugh Ingle's small mining operation near Hawthorne, Nevada. They were accompanied by Hugh and by Russ Fields to the site, which is a small, high-grade underground mining operation. After touring the area immediately surrounding the mine, both Rahall and Vucanovich were taken on a tour of the underground mine workings.

The purpose of the underground mine visit was to acquaint the subcommittee with small mining operations and have an opportunity to point out the difficulties that small-scale miners face in their efforts to produce minerals. It was hoped that the difference in perspective between an onsite visit to a large mining operation versus a very small underground mine would help to illustrate the impact that even minor changes to the 1872 Mining Law would have on small-scale miners. In this respect, the tour was reported to be an unqualified success.

However, it remains to be seen if the combination of the hearing testimony and the small-mine visit will have any appreciable effect on Rep. Rahall's approach to mining law reform. In any event, Nevada miners gave it their best shot.

22
Strong Opposition Shown Towards Senator Bumpers' S.433

DURING A RECENT HEARING (June 1991) in Washington, D.C., on Sen. Dale Bumpers' S. 433 (the so-called "Mining Law Reform Act of 1991"), a coalition of Westerners served notice that they are prepared to fight any effort to repeal the 1872 Mining Law, as amended. The opponents appearing at the subcommittee hearing included Sen. Harry Reid (D-NV), Sen. Richard Bryan (D-NV), and Rep. Barbara Vucanovich (R-NV), who all said that the existing mining law is vital to the health of their regional economy.

Sen. Reid countered Bumpers' claim that the mining industry takes $4 billion worth of minerals from federal lands each year without paying any royalty or providing any return to the general public.

"The royalties returned to the people of this country are immeasurable," Reid said. "Perhaps the greatest royalty to the people is national security.... We all know the dangers of being dependent on foreign oil. Let's not become dependent on foreign minerals."

Sen. Bryan noted that mining is a critical element of the Nevada economy.

"I know that some abuses of the mining law for non-mining purposes have occurred, and I believe there is a broad consensus within the industry as well as outside to prevent such abuses," Bryan said. "But for those of us who represent public land states—Nevada is comprised of nearly 87 percent federally owned land—it is critical that mining reform not spell the demise of our mining industry."

Both Sen. Bryan and Sen. Reid expressed support for a measure introduced by Sen. Conrad Burns (R-MT), S. 785, that calls for a commission to study the regulations and laws governing mining on the public lands.

Rep. Vucanovich stressed that Bumpers' S. 433 would likely put many miners out of business by imposing a 5% royalty on the gross production of hardrock minerals. This royalty would have a "severe impact" on Nevada's mining industry particularly when coupled with the state's 5% tax on net production, she said.

"Taxing profits does little to drive miners out of business, but taxing the gross proceeds just makes an already risky business that much more volatile," Vucanovich said.

Sen. Alan Simpson (R-WY) called Bumpers' bill a "staff-driven projectile" and he said he is tired of hearing about the "subsidization of the West."

"I've been here 13 years, and we've subsidized everybody else, including the corn growers in Iowa and Arkansas," Simpson said.

Sen. Dennis DeConcini (D-AZ) testified that the 1872 Mining Law is not a giveaway and said that S. 433 would jeopardize the mining industry in the West.

Sen. Tim Wirth (D-CO), who is a member of the subcommittee, said the mining law has served the country well, and that wholesale revision of the law was not necessary. He also voiced skepticism over the appropriateness of a royalty on hardrock minerals.

Subcommittee Chairman Jeff Bingaman (D-NM) also stated that a "wholesale revision of the mining law is an inappropriate way to address these problems." Bingaman added that the Subcommittee on Mineral Resources Development and Production will be conducting field hearings on the mining law bills and the issues involved.

BLM Director Cy Jamison said the Administration "strongly opposes S. 433." He emphasized that the BLM's planning and reclamation requirements, together with the agency's surface management, effectively addressed the concerns being expressed with respect to the mining law.

23
Is 'Activism' a Dirty Word for Miners?

Editor's Note: The following item is reprinted from the Perspective/Opinion page of the March/April 1992 issue of *Mining World News*.

ITEM: A *Sacramento Bee* reporter wins the Pulitzer Prize for public service for an examination "of environmental threats and damage to the Sierra Nevada." Among those "threats" is an article on pollution caused by mining. That same journalist, now honored by his peers, wrote a series of articles which set a new low in mine-bashing.

ITEM: The Secretary of Interior recommends substantial cuts in the research budgets of the Bureau of Mines and U.S. Geological Survey. An assistant secretary tells a bureau administrator the Interior Secretary feels justified in making the cuts because no one employed by mining bothered to protest.

ITEM: A state wildlife official, the General Accounting Office, and numerous experts have repeatedly offered evidence to Congress and the Environmental Protection Agency that cyanide can be properly and safely used in mining processes with no consequent harm to wildlife. Yet politicians, the EPA, and the news media repeatedly raise the issue without fear of any serious public challenge by the chemical or mining industries.

What is it going to take to fuel the fires of activism in mining? Apparently, all of the preceding items aren't sufficient to get mining professionals motivated. Our numerous opinion pieces (in the *MWN*) urging people to take action apparently are falling on deaf ears.

This is the age of activists against AIDS, the pro-lifers, the National Organization for Women, the Sierra Club, the Mineral Policy Center, the United Auto Workers, the political action committee, the League of Women Voters, and People for the Ethical Treatment of Animals. In essence, a lab rat has more political clout with the executive branch or Congress than a mine, the mining industry or your livelihood. Yet, mining professionals hire high-powered lobbyists or big-bucks public relations firms, pay dues and prefer to let the local or national mining association or organization handle the mess.

Executives running multi-million-dollar mining operations can't be bothered with well-organized, radical and highly visible lobbying and media campaigns. They're much too busy—or, perhaps, too fearful of regulatory retaliation—to engage in the kind of aggressive, forthright, dramatic activism practiced by their opponents.

But what happens when a project is killed or millions of dollars in legal fees and environmental mitigation are spent because someone else did get involved and challenged your company's right to mine?

Hate the negative, anti-mining headlines? But, are you too busy in the field or afraid to visit the editorial board of your local metropolitan newspaper and explain the benefits of the mining industry? Does your company or association water down the one valuable communications tool at its disposal—a newsletter or corporate report—for fear of being too controversial or, worse yet, proactive?

Is your company moving operations to Latin America or the Commonwealth of Independent States because the U.S. political and regulatory environment doesn't favor mining?

Can't you fax a letter to the Secretary of the Interior, urging him to promote the interests of the domestic mining industry by sufficiently funding mining research budgets? Can't you tell a political candidate that you only give campaign donations to those who are willing to seriously consider domestic mining's interests?

Are you still donating money or paying dues to a Sierra Club, a National Wildlife Federal, or any charity or cause to be a "good neighbor" while a People for the West! organization or a mining association education campaign is crying for funds?

Shouldn't you unite with other mining professionals and your suppliers, get in the trenches and battle to save your industry, promote valuable research, and fight for your right to earn a living in this country?

Is *activism* a forgotten word in your vocabulary?

Special Note: The *MWN* editorial appearing above is directed primarily toward the corporate executive, CEO, mine manager or other mining industry professional; but aren't most of us (small-scale miners, mine workers, suppliers, mine equipment manufacturers, prospectors, recreationists, dredgers, etc.) afflicted with a similar reluctance to get involved in the fray?

We are (metaphorically speaking) all in the same boat together, have our livelihoods at stake, and are being subjected to the same vicious assaults by anti-mining activists. Why is it, then, that so very few of us are actively involved in doing something about this madness? Are we waiting for all mining activity in America to be eliminated before we finally have had enough and become willing to respond to the insanity directed against us?

Procrastination yields absolutely nothing; things almost always get worse if you put them off. In this case, the end result is guaranteed. If we don't all get involved now, there may be nothing left that is worth fighting for.

PART III

Defending the 1872 Mining Law

"Here we go again! Just when miners were beginning to feel that the assault on America's mining community by the U.S. Congress had reached its peak, several more politicians decided to jump on the bandwagon. In an apparent attempt to appease the radical environmentalists...a few more members of the Congress have decided to introduce their own version of mining law "reform.""

24
The History of U.S. Mining Law

THE ORIGINAL BASIS FOR U.S. mining laws evolved from English and Spanish concepts developed in the 16th Century. Several of the early American colonial charters provided for grants of mineral lands to colonists in a manner similar to the initial Crown land grants to the first settlers. Exclusive rights to precious metals were usually reserved for the English sovereign. This system regulated mineral rights in the original 13 colonies and much of the eastern U.S. until the mid-1800s.

In the Western region, however, most mining rules and traditions were based upon the Spanish Royal Code of 1783. This code provided the general rules for acquiring mining rights and for settling conflicts between miners. Because there was no U.S. mining law prior to the California Gold Rush in 1849, these customs were observed by miners throughout the early West.

Until this time period, Congress had only passed several temporary sales and leasing acts dealing with gold, silver, iron and lead, with the provisions of these various acts being administered by the U.S. War Department. So, when the 49ers began locating mining claims they legally had no rights to the gold they mined.

By the early 1860s, the development of the Comstock Lode at Virginia City, Nevada, again pointed out the need for security of title to mineral discoveries and the right to develop and produce minerals. Since Eastern capital was financing a large portion of the development on both the Comstock Lode and the California Mother Lode District, this matter quickly became a major political issue in the U.S. Congress.

Further, congressional policies concerning public domain lands in the period from 1865 through 1885 focused on means to encourage westward migration of settlers to develop the Far West. As part of this overall plan, Congress enacted a series of laws that included the several Homestead acts, agricultural entry, U.S. soldier compensation acts, and legislation designed to encourage mineral exploration and development.

The first U.S. mining law, called the Lode Law of 1866, was

passed by Congress on July 26, 1866. The law authorized procedures for entry and location of lode claims and contained specifications for performance of assessment work and the patenting of lode claims. This was followed by enactment of the Placer Act on July 9, 1870, which covered the entry and location of placer claims, location by legal land survey description, and procedures for obtaining mineral patents. These two laws were then combined with various amendments into the General Mining Law of 1872 on May 10, 1872.

U.S. Sen. William M. Stewart of Nevada played an important role in the eventual construction and enactment of these mining laws which, with numerous amendments, still provide the basis for mineral development in the United States today.

The overall scope and application of the 1872 Mining Law has changed considerably since the law's passage in the nineteenth century. Succeeding congressional amendments and additions to the legislation, the evolution of mining case law, the enactment of administrative laws, and the formulation and application of federal, state and local environmental regulations have dramatically changed the mining laws in conformance with the changing environmental and natural resource policies of the country.

For example, the Pickett Act in 1910 effectively withdrew federal lands from entry for coal, oil, gas and phosphates under the mining law, and the Potash Leasing Act of 1917 removed potash. The Mineral Leasing Act of 1920 removed oil, gas, coal, sodium, phosphates, oil shale and several other minerals from location by mining claims and stipulated that they must be leased from the government. Finally, the Surface Resource Act of 1955 removed common varieties of sand, gravel, stone, pumice, pumicite, cinders, clays and other industrial minerals, which were then made salable and leasable commodities by competitive bid procedures.

Environmental concerns were addressed by the Clean Air Act in 1963, the National Environmental Policy Act of 1969, the Clean Water Act in 1972, the Resource Conservation and Recovery Act in 1976, and the Federal Land Policy and Management Act of 1976. All of this legislation effectively amended and updated the 1872 Mining Law to provide for other concerns that had not been written into the original legislation.

In addition, numerous federal agency regulations have been formulated and implemented to cover every activity now authorized under the mining laws. Wherever mining operations are conducted in the nation, state and local laws, rules and regulations are also applied and enforced.

When it was first enacted, the 1872 Mining Law provided the necessary conditions for the creation and maintenance of a large and productive mining industry in the United States. With subsequent amendments and additions over the years, the law has worked very well for over one hundred years by providing reasonable incentives, an efficient framework, and a dependable basis for the nation's mineral resource exploration and development. The law's structure has accommodated numerous changes in our resource development policies, implementation of environmental controls, land use planning and policy decisions, and the technological advancement of the country's mining industry.

The federal laws covering mining and minerals are contained in Title 30 of the United States Code (30 U.S.C.1-1811). Individual states have enacted statutes that define approved procedures for the location and maintenance of mining claims, and the requirements vary somewhat from one state to another.

25
Senator Bumpers' S.1126 and Potential Impacts of Similar Legislation

CONGRESSIONAL LEGISLATION that would have effectively repealed the Mining Law of 1872 was introduced in Congress on June 1, 1989, by Sen. Dale Bumpers (D-AR), and a hearing was held on the measure on June 7. Bumpers' S. 1126 bill contains at least eight major changes to the existing minerals legislation as well as repealing the provisions of the basic mining law and all amendments thereto. The horror story envisioned by Sen. Bumpers illustrates just how critical the mining law issues have become.

Fortunately, S. 1126 did not receive serious consideration in the 1989 Congress, but it is highly likely that it will resurface in some form during the 1990 session. In addition, the House of Representatives also discussed several proposed changes in the mining laws during 1989, although no formal legislation was introduced. It is also likely that these efforts will be considered again this year. A considerable amount of pressure to abolish the 1872 Mining Law is being exerted by extremist elements in the environmental movement and, unfortunately, a number of our congressional leaders have chosen to back this effort.

Since many of the same or similar proposals contained in Bumpers' bill will again be placed before the Congress, an analysis of this legislation and its potential impacts on miners, and mining in general, is particularly relevant. Comments on specific sections of the bill are provided in Appendix C.

General Comments

There is a considerable discrepancy in the definition of "hardrock minerals" contained in Title I, Section 102, of the bill. Although the bill definition "means any mineral of a kind which on June 6, 1989, was subject to location under the Mining Law of 1872," which would include lode, placer and specific industrial mineral deposits, the definition of "hardrock minerals" in both case law and as used in the current mining laws would only apply to mineralized ores in lode deposits. Since the Mining Law of 1872 would still apply to existing claims on the date of enactment of this bill, a legal interpretation of the term "hardrock minerals" would most likely follow the current legal precedent. In any case, it seems this conflict in terms would have to be settled in the courts. Because of its critical importance, the process could go on for many years.

The basic language in S. 1126 is penalizing in nature: numerous strict compliance requirements with multiple provisions for fines and automatic invalidation of claims "by operation of law" (read: at the discretion of, or for just about any purported violation). The penalty provisions are spelled out exactly, with no mention of due process or provisions for appeal, yet most other sections dealing with items of major importance to miners are subject to vague interpretations and a wide-open invitation to formulate additional stringent regulatory standards.

Most importantly, the overall thrust of the bill and its approach to mining activities in general is extremely negative. The content of the legislation illustrates a definite anti-mining bias and an extremist environmentalist nature on the part of those who drafted and supported such a proposal. This makes it highly likely that any future legislative measures coming before the U.S. Congress on mining issues will utilize much the same approach and exhibit a similar bias against the mining industry.

Projected Impacts of This Type of Legislation

This type of bill would ensure the elimination of small miners, prospectors, recreational prospecting, gold dredgers and small equipment manufacturers. The onerous burden of excessive costs, expenditure requirements, and federal royalties would preclude the development of new small mining operations, and it would sound the death knell for existing small mining operations within just a few years. These costs would also severely impact the larger mines and the industry as a whole, producing an almost immediate decline in mineral production and a longer-term permanent decline in mining throughout the nation.

The mandatory bonding requirement for *all* mining activities would automatically void the majority of existing mining claims and preclude the development of new mining operations under the Act. At best, bonding for mining operations is very difficult to obtain under existing laws. Under this type of legislation, bonding would be impossible to obtain in the vast majority of cases—except perhaps the very large, profitable existing mines. Costs of available bonding would also increase to very high levels.

The impact of this type of legislation on U.S. economic activity would be devastating: (1) mines would close, producing employment losses and a dramatic drop in local economic activity in the communities where the mines are located; (2) the majority of mining claims would be dropped, producing a sharp drop in revenues to federal land management agencies that would, in turn, have to be subsidized by the government (higher federal taxes); (3) mineral production would fall off sharply, producing a dramatic rise in mineral import reliance on foreign sources and increasing the foreign trade deficit; (4) many mining suppliers (equipment manufacturers, retail outlets, publications and peripheral businesses) would be forced out of business, thereby reducing economic activity and producing more unemployment; (5) the reduction in supplies of mineral commodities would put upward pressure on mineral prices, thereby inflating the cost of consumer goods, and (6) the ripple effect on the U.S. economy as a whole would be very significant and possibly quite severe.

If S. 1126 had been enacted by Congress, the legal implications would have been mind-boggling. The current mining case law has evolved over a long period of time, and most of it would suddenly be questionable. For existing claims, legal ramifications and the resolution of same in the courts would certainly involve extensive litigation, exorbitant costs and a considerable amount of time—perhaps decades. For new claims located under the Act, relevant case law would have to be established from scratch. This would take a very long time and involve considerable expense, and the process would undergo constant amending and revision as new regulations came into effect—let alone the impact that ongoing congressional actions would create. The bill is a disaster from the legal standpoint.

Most unfortunately, miners and the mining industry can fully expect that similar legislation will be proposed in the U.S. Congress in the very near future. The prospect is horrifying.

26
HR 3866: Rahall's 'Mineral Exploration and Development Act of 1990'

CONGRESSMAN NICK RAHALL (D-WV), chairman of the House Mining and Natural Resources Subcommittee, in January introduced legislation in the House of Representatives that would repeal the Mining Law of 1872 and replace it with the "Mineral Exploration and Development Act of 1990." In his introduction of H.R. 3866, Rahall stated:

> Today I am introducing legislation to revise the Mining Law of 1872. This bill is aimed at providing a focal point for debate in the House of Representatives, after a hiatus spanning more than a decade, on some very pressing issues facing the future of mineral exploration and development on public domain lands in this country.

Rahall said that the 1872 Mining Law survives as the last vestige of the nineteenth-century western settlement measures and that some of the current provisions of the laws thwart efficient mineral exploration and development. He asserted that "nothing short of legislation can fix this situation."

Overview and Background

Rep. Rahall described H.R. 3866 as a mining claim bill that is based on the principles of access to public domain lands and the right of self-initiation. The measure would eliminate the present concept of "discovery" because, he said, discovery "is an illusory concept that is grounded in a quagmire of judicial and administrative pitfalls, and fraught with time-consuming and expensive legal proceedings."

The legislation provides that once a mining claim is located and recorded, possessory rights are protected so long as there is compliance with rental, diligent development, and filing requirements established by the bill. H.R. 3866 does not propose any type of federal production royalty. However, it eliminates the distinction between lode and placer claims, eliminates extralateral rights, eliminates mineral patents, and requires that mineral development be subjected to surface management regulations and the land use planning process.

Rahall said in his remarks on rental fees and production royalties:

> The proposed rental rate is not set so as to burden small prospectors, yet at the same time, it would provide some return to the public for the

use of its lands. In the same sense, the proposed diligent development requirements of the legislation are set low enough during the first few years after the location and recordation of a mining claim so as to facilitate, and not hinder, mineral prospecting and exploration activities.... As such, this bill is not aimed at extracting revenue from holders of mining claims. This should not be the purpose of mining legislation.... I am not proposing that a production royalty be imposed on minerals produced from mining claims.... I cannot in good conscience support the imposition of royalties on hardrock mineral production.

In his closing remarks Rahall stated:

This legislation has been a year in the making, and we have taken great pains to solicit input from anyone who cared to work with us. However, I do not purport to have devised a perfect piece of legislation. Be that as it may, this proposal is representative of some very basic tenets that I believe should be discussed as part of any consideration of legislation in this area.... As I stated earlier, the purpose of this legislation is to begin once again consideration of the need to improve upon the type of regime set forth by the Mining Law of 1872. I am extremely open to comments from all interested parties, and look forward to our future deliberations.

A careful examination of Representative Rahall's H.R. 3866 reveals that it is a more balanced and reasonable approach than that taken by Sen. Bumpers in his S. 1126, "The Mining Law of 1989." Even though Rahall's proposal contains a number of significant problem areas, there is still no comparison with Bumpers' punitive and damaging approach to revising the Mining Law. A brief description and analysis of each section in Rep. Rahall's H.R. 3866 is provided in Appendix D of this volume (page 211). Below are comments on the most significant changes and additions proposed.

Summary of Most Significant Impacts

On the positive side, the bill preserves the right of self-initiation and provides for protection of possessory rights. It also eliminates the "prudent man" and "marketability" standards now applicable under current law, so claims cannot be invalidated for lack of "discovery" under present law. The bill also does not include a provision for federal production royalties or extremely onerous and punitive measures such as those contained in Sen. Bumpers' 1989 proposal. In addition, the bill outlines fairly comprehensive procedures for contesting and appealing adverse findings and civil penalties.

In some cases, the elimination of extralateral rights and mineral patents may prove to be problematic for some miners.

The requirement for locating 40-acre claims by legal subdivision may prove to be impossible in areas where land situations are tight (existing private property, restricted areas, existing mining claims), or where some extreme adjustments have been made in the public land survey system. The conversion of existing mining claims to the new system in active mining districts would be a nightmare, and the relocation of certain claims would likely move some of them from proven mineralized areas. Provisions are not made for fractional claims, and this would be absolutely essential for the conversion of many existing claims to the new system. Because of overlapping claims, mineral patents, existing claims and private property, certain mineralized areas could never be located under this requirement.

The *minimum* rental fee would increase holding costs by $60 per claim for 40-acre claims ($30 for 20 acres) annually. This would be a significant expense on larger claim groups, especially for small miners and prospectors. Keep in mind that this fee may be higher, and it would certainly increase over time. The $5 per acre fee on smaller mining operations, such as placer dredging, would increase cash costs by $200 per year for a 40-acre claim, which would produce a significant financial impact on small miners. These "user fees" would particularly impact large exploration projects. Recent estimates by federal agencies show that about 85% of the mining claims in the U.S. are held by small miners and prospectors, so they would bear the major financial burden. Many claims would be dropped as a result.

The *minimum* annual diligent expenditure requirements would effectively quadruple over the first five years under this Act. They would be 8 times as much over the next five years, increasing to 16 times in the next five years and to 32 times the current expenditure requirement thereafter. This increase in expenditures would produce a major impact on mineral exploration properties and result in the forfeiture of large numbers of mining claims. Exploration is a risky and expensive business, and these costs would act to discourage future mineral exploration and development. In addition, the expenditures would be required during the first year the claim is located, which would add considerably to the initial staking, filing and recording costs.

The record-keeping requirements to absolutely prove the diligent expenditures were made would mean that individuals performing the work would have to have witnessed documents, pictures, receipts and reports verifying each and every expenditure in personal labor and associated costs. This might be quite difficult for individual prospectors and miners, especially when their records are audited by the government. However, the provisions in the bill to allow additional types of work for qualified expenditures will be helpful in reflecting the actual costs of exploration work and the cost of meeting regulatory requirements.

Depending upon the approach used in formulating the required new rules, standards and regulations, compliance with same could be anywhere from fairly straightforward to very difficult. These new requirements would be a critical factor in the implementation and workability of the new laws.

The Act's proposed inclusion of mined land reclamation and environmental concerns in mining law is not really necessary, since these issues are already addressed in a multitude of other federal and state statutes and regulations. The potential here is that any new environmental standards and reclamation requirements formulated under this bill's requirements may be much stricter and more unrealistic than those already in force. On the other hand, a reasonable approach to formulating new regulations in this area that specifically address mining issues might reduce some of the current confusion and duplication. However, this is rather unlikely given the present trend towards overkill and the provision allowing individual states to enact measures that are in compliance with federal requirements but are much stricter in addressing environmental concerns.

The elimination of uncommon varieties of industrial minerals from location by mining claims is another area of concern. Many of these deposits would not be economically feasible to operate if they were placed under a salable and leasable system because of the additional associated costs.

The transfer of administrative authority for minerals management to the U.S. Forest Service would undoubtedly lead to much stricter regulatory requirements on National Forest System lands than on BLM lands. In effect, this would authorize a dual and different system of minerals management.

By allowing the conversion of existing mining claims located under the 1872 Mining Law to the new system without applying the relocation by legal subdivision requirement, the bill does attempt to protect prior rights of mining claimants under existing law. However, the proposed rental fees and increased diligent expenditure requirements would still increase the cost of maintaining claims significantly. In addition, no provision is made for relocation of existing claims under the new system, and most of these claims could not be located under the new 40-acre requirement.

27
Rep. Rahall Introduces New Mining Law Bill

AS EXPECTED, REP. NICK RAHALL (D-WV) has finally come out with his revised version of H.R. 3866, which was first introduced in January 1990 and was the subject of a public hearing last September 6 (see "An Overview and Analysis of Rahall's H.R. 3866..." March 1990 *CMJ*, and "Miners Show Strength and Unity at Mining Law Hearings" in October 1990 *CMJ*).

On February 6, 1991, Rep. Rahall introduced H.R. 918, the "Mineral Exploration and Development Act of 1991." However, the contents of this new bill were totally unexpected: The legislative proposal has grown by from 50% to 60% (31 pages to 46 pages, plus more on each page); the annual rental fees are still a *minimum* of $1.50 *per acre* per year for mining claims and a *minimum* of $5 *per acre* per year for claims under a plan of operations; the "diligence" (annual assessment work) requirement is still *$20 per acre* in years 1-5, *$40* per acre in years 6-10, *$80* per acre in years 11-15, and *$160* per acre in year 16 and thereafter; penalties for violations of claim maintenance rules increased from *$500 per violation per day* to *$5,000 per violation per day*; penalties for failure to comply with surface management requirements increased from *$1,000 per violation per day* to *$5,000* per violation per day, and *user fees* charged for the administration and enforcement of the Act *can be set at any level*.

All existing federal, state and local laws, rules, regulations, fees, taxes and reclamation and environmental standards that are not in conflict with H.R. 918 will still apply, so the requirements of this bill will be added onto most existing fees, permits, taxes and regulations.

The bill makes all mining activity subject to the land use planning process, so mining can be disallowed whenever and wherever it is in conflict with land use plans (both present and future). Mining activities would be subject to the bill's new environmental and reclamation standards, in addition to existing environmental and reclamation requirements under federal, state and local laws, regulations and ordinances.

Under the bill, a mining plan of operations will be required for all mining activities that produce almost any type of land disturbance, including use of any mechanized equipment and suction dredges. In other words, the 5-acre rule has been eliminated and even small operations must now have an approved plan of operations. Minimal surface disturbances (drilling with a track rig, digging hand holes) may be conducted under a notice, but even these negligible disturbances must be reclaimed. An "adequate financial guarantee" (bonding or surety) will be required for both notices and plans of operations, and it must be posted before any activity takes place.

These are just a few highlights from a complex and detailed piece of legislation. A complete section-by-section explanation and analysis of Rahall's H.R. 918 will be published in the April issue of the *CMJ*.

The political implications of this bill are also noteworthy. Rep. Bruce Vento (D-MN) and Rep. George Miller (D-CA) are also sponsors of Rahall's bill, and Rep. Miller is now the acting chairman of the House Committee on Interior and Insular Affairs. Congressional staffers say the bill is slated for quick action.

28
An Analysis of S.433, Senator Bumpers' Mining Law Reform Bill

Editor's Note: See related article, "Senator Bumpers' S. 1126 and Potential Impacts of Similar Legislation," on page 131 of this volume.

SENATOR DALE BUMPERS (D-AR) introduced S. 433, the "Mining Law Reform Act of 1991," on February 20, 1991. The bill is a revised and expanded version of his S. 1126, introduced on June 1, 1989. The new bill contains most of the provisions included in the earlier proposal, but the majority of them have undergone rather extensive revision and, in some cases, replacement. Most importantly, the new legislation contains a major expansion of environmental and reclamation requirements, which are now listed under Title II and III.

One of the most disturbing aspects of the mining law legislative "reform" being proposed by Sen. Bumpers is the retention of his punitive approach to the subject. As in S. 1126, Bumpers has designed S. 433 to effectively shut down mining in the U.S. or, failing that, to impose huge fines and jail sentences on all miners who (heaven forbid) might occasionally make a mistake in following his new rules.

As expected, almost all of the testimony provided by mining representatives at the hearings held in 1989 and 1990 was, for all intents and purposes, ignored. However, most of the extremist statements made by environmental groups in their testimony have definitely shown up in S. 433—most particularly by the specific authorization of perpetual litigation by "any person" who wishes to file civil action at any time for any reason, and also by the provision for perpetual and unlimited liabilities for those who might still be foolish enough to try to mine minerals in this country.

There is a rather positive aspect to this bill: It is so bad that it might fail all by itself. It is easy to oppose: We can again afford to just say no to the entire package rather than waste our time in efforts directed toward changing specific provisions. In other words, Sen. Bumpers' S. 433 can't be fixed.

However, political reality must also be considered. Sen. Bumpers does have a lot of political clout. He has succeeded in getting seven other senators to cosponsor the bill. Quite a number of senators would not be terribly concerned about efforts to change the mining law, and S. 433 will have the complete backing of most of the nation's environmental groups. In other words, even as bad as the bill is, it still has a fairly even chance of being enacted by Congress.

There is almost no chance at all for an effective challenge of this bill at the committee level—it will undoubtedly come down to a vote on the floor of the U.S. Senate. Therefore, opposition to the measure must be communicated to each and every member of the Senate. The potential seriousness of the situation cannot be overemphasized, and this can be most effectively illustrated by listing the major projected impacts if S. 433 is passed.

Major Impacts of S. 433

1. Title II of S. 433 (which is virtually identical to Title II in Rahall's H.R. 918) requires the imposition of excessively stringent environmental and reclamation standards, and all of these requirements would automatically apply to all existing mining claims and mining operations as of the effective date of the bill (in this case, one year after the date of enactment). There are no time constraints on retroactivity and applicability of Title II, so it would make every claimholder responsible for all surface disturbances and environmental problems that exist on their claims on the effective date of the Act. In other words, it would create unlimited liability for all claimholders.

2. The onerous burdens of excessive regulation and major cost increases would ensure the elimination of all small-scale miners, prospectors, recreational prospecting, gold dredging, and small equipment manufacturers within just a few years. These burdens would also severely impact the large mining operations and the industry overall in a very short period of time.

3. In the place of assessment work and/or development work, S. 433 requires that only cash payments be made to maintain a mining claim. Claimants must pay a $100 fee to record each claim, make a payment of *$5 per acre* in each of the first through fifth years, pay *$10 per acre* in years 6-10, pay *$15 per acre* in years 11-15, pay *$20 per acre* in year 16 and each year thereafter, pay a federal *5% gross production royalty* on all mineral production, pay an annual surface use fee of *not less than $5 per acre* for mill sites, etc., and pay additional *user fees* for administration of this Act.

4. The bill provides for civil penalties of not more than *$1,000 per day per violation* for failure to comply with provisions of the Act, and *criminal penalties of not more than $50,000* or *imprisonment for not more than 5 years* or *both fine and imprisonment, for each day of any violation* by any claimholder who "knowingly and willfully" violates any provision of this Act.

5. Provides for "Citizen Suits" that specifically authorize "any person having an interest which is or may be adversely affected" to file suit in a civil action against the United States or against any other person who "is alleged to be in violation of any rule, regulation, order or permit issued pursuant to this title" (Title II). This provision opens up any exploration or development project to perpetual litigation in the courts, at any time and for any reason. No sane person or company would invest time and money in a mining project with this type of unlimited risk; you could be shut down and hauled into court at any time.

6. S. 433 requires that a mining Plan of Operations be approved prior to making just about any type of land disturbance, including any use of mechanized equipment, suction dredges, off-road motor vehicles, etc. This means that even extremely small exploration projects and mining operations would be subjected to Environmental Impact Statements, public hearings, expensive permits, mandatory bonding or surety, and land use planning.

7. *All other existing laws, rules and regulations will still be in force.* Sec. 404, "Savings Clause" of S. 433 states:

> Nothing in this Act shall be construed to repeal or modify an existing law that prohibits or restricts the application of the Mining Law of 1872 to public domain lands where such law provides for greater protection of such lands than the provisions of this Act. Nothing in this Act shall be construed to effect preemption of state laws or regulations where such laws or regulations provide for greater protection of the environment, including state laws or regulations requiring higher bonding amounts or more stringent reclamation standards than this Act, *except for preemption of state and local location and filing requirements pursuant to Section 101(c) of this Act.*

In other words, all existing permits, fees, taxes and state requirements will still apply, *in addition to the requirements of this Act.*

Summary

The preceding list outlines only a few of the major impacts of this legislation if it is, in fact, enacted by Congress. It was felt that this approach would provide a better understanding of the overall devastating effect S. 433 would have on miners, rather than just providing a detailed description of each section in a very complex legislative proposal.

The cumulative impact of S. 433 on the mining industry and U.S. economic activity would be dramatic: (1) Mines would close, producing employment losses and a sharp drop in local economic activity in those areas where mines are located; (2) the majority of mining claims would be dropped, producing a consequent loss of economic activity and a permanent drop in potential future mineral production; (3) mineral production will fall off sharply, producing a dramatic rise in mineral imports and an increased reliance on foreign sources of supply; (4) many mining suppliers (equipment manufacturers, retail outlets, publications, service industries and peripheral businesses) would be forced out of business, thereby reducing economic activity and producing more unemployment; (5) reduced availability of domestic mineral supplies would put upward pressure on mineral prices, thereby inflating the cost of consumer goods and housing, and (6) the ripple effect of higher costs, unemployment and reduced economic activity on the U.S. economy as a whole would be very significant and possibly quite severe.

Of course, there would eventually be no mining industry in the U.S. at all, and all minerals production would move offshore.

29
An Analysis of HR 918, Rep. Rahall's New Mining Law Bill

EARLY THIS YEAR, REP. NICK RAHALL (D-WV) introduced H.R. 918, the "Mineral Exploration and Development Act of 1991." H.R. 918 is a revised and greatly expanded version of his H.R. 3866 introduced in January 1990. The new legislation contains most of the provisions included in last year's bill, particularly those requiring much higher mining claim holding costs, as well as a major expansion of environmental and reclamation requirements.

One of the most disturbing aspects of the mining law legislation is the fact that the comprehensive and factual testimony offered by mining representatives during a previous hearing, held in Washington, D.C., on September 6, 1990, has been ignored. Almost all of the provisions objected to by 26 miners who appeared at the hearing have survived intact, while a number of the more onerous provisions have increased in severity. This would indicate that the potential for addressing these major issues in future committee hearings and obtaining any constructive changes is almost nil.

As a result, miners can no longer afford to maintain a proactive stance in addressing specific provisions of the bill and must instead adopt a position of total opposition to the entire legislative measure. Because of the politics involved, this opposition must be expanded to include individual contact with all members of the U.S. Senate and House of Representatives. The importance and seriousness of H.R. 918 cannot be overemphasized, and this is best illustrated by listing the major impacts of the bill.

Major Impacts of H.R. 918

1. As written, H.R. 918 would create virtually unlimited liabilities for the nation's minerals producers and, as a result, would signal the permanent decline of U.S. minerals production.

2. The bill would, as of the effective date, almost immediately eliminate the participation of most small-scale miners and prospectors in both existing and future mineral exploration and development.

3. Title II of H.R. 918 calls for the implementation of excessively stringent environmental and reclamation standards, and these re-

quirements would apply to all existing mining claims and mining operations as of the effective date of the bill. Without any time constraints on retroactivity and applicability, this title automatically makes every claimholder responsible for all surface disturbances and environmental considerations that exist on their claims on the effective date of the bill. As this title reads at present, it is, in effect, Superfund II.

4. While the bill states that it provides security of title to mining claims, it effectively eliminates security of title to mineral deposits. Mining can be disallowed at the discretion of land management agencies and/or by objections by members of the general public.

5. The measure specifically provides for the filing of lawsuits by any person for just about any reason, thereby opening up mining operations to the possibility of perpetual attack and the very real potential for being shut down at any time. No reasonable person or company would invest exploration or development capital in a mining project with this type of unlimited risk.

6. The bill requires that a mining Plan of Operations be approved prior to even minimal land disturbances by exploration activities or mining operations, including minor trenching and suction dredging, thereby subjecting small exploration projects and dredging activity to provisions in the National Environmental Policy Act (NEPA process), public hearings, expensive permits, long time delays, bonding or surety, and the land use planning process. While negligible land disturbances *may* be allowed under a Notice, bonding is still required prior to initiating any activity. This provision alone would effectively kill most small exploration projects and small mining operations in the United States.

7. The major increase in annual "diligence" (assessment work) requirements would automatically elevate most annual assessment work activity to the Plan of Operations level over time, especially after the first 5 years. Even relatively minor work on the ground would require the filing of a Notice or Plan, both of which require bonding or surety. Therefore, claimholders would find themselves in the position of having to obtain a bond and go through the lengthy and costly Plan of Operations process just to meet their annual diligence requirements.

8. The legislation stipulates that all existing federal, state and

local laws, rules, regulations, fees, taxes, ordinances and reclamation and environmental requirements will still apply to all existing and new mining claims, if they are not in conflict with the provisions of H.R. 918. This means that the requirements of this bill will be added onto most of the existing permit requirements, fee structures, taxes, and environmental and reclamation standards.

9. Most of the basic claim maintenance costs are the same as they were in last year's proposal: Annual rental fees are a *minimum* of $1.50 per acre per year for mining claims and a *minimum* of $5 per acre per year for claims under a Plan of Operations (and most claims would eventually fall in this category); the diligence (assessment work) requirement is $20 per acre per year in years 1-5 (including the first year), $40 per acre in years 6-10, $80 per acre in years 11-15, and $160 per acre in year 16 and all years thereafter. In addition, there is a provision for *user* fees to be charged for administration and enforcement of the Act (with no upper limit). Adding these increased costs on top of the existing cost structure would be prohibitive to most small miners, even if they could afford to comply with the increased costs associated with regulatory and legal requirements.

10. Under the enforcement requirements, the penalties for violations of claim maintenance rules have increased from *$500 per violation per day* last year to *$5,000 per violation per day* under H.R. 918, and the penalties for failure to comply with surface management requirements have increased from *$1,000 per violation per day* to *$5,000 per violation per day* under the current bill. Under Title II of H.R. 918, as it is currently written, almost all existing mining claims would automatically be in violation on the effective date of the bill.

11. The requirement to prepare land use plans with the express purpose of protecting nonmineral values makes it likely that mining will be disallowed from large areas of nonsensitive public domain lands, at the discretion of land management agency personnel, in addition to the withdrawals of huge tracts of land from mineral entry that already exist.

12. H.R. 918 also contains provisions to eliminate mineral patents and extralateral rights and removes mineral materials from location with mining claims—including the so-called uncommon varieties.

13. The bill also calls for the transfer of minerals management on U.S. Forest Service lands from the Department of the Interior to the

Department of Agriculture which, in effect, authorizes a dual system of minerals management on the public lands.

Overview and Summary

The preceding list of the major impacts of H.R. 918 was presented to provide a better understanding of the cumulative effects of the legislation if it is enacted by Congress, rather than just listing the requirements of each title and section in a very complex and detailed piece of legislation. Overall, the basic foundation of this bill is very similar to that of last year's H.R. 3866, particularly in reference to claim location and maintenance requirements and provisions for the conversion of existing claims to the new system. However, the major thrust of this so-called mining law reform measure has changed dramatically: H.R. 918 is now much more of an environmental bill and much less of a mining measure. In fact, well over half of the bill's contents is devoted to environmental legislation.

If enacted, H.R. 918 would effectively abolish the 1872 Mining Law and replace it with an environmentalist agenda; the environmentalists would ultimately have complete control of all mining activities in the United States. In the final analysis, the "Mineral Exploration and Development Act of 1991" is a contradiction in both its supposed intended purpose and terminology: it would put mining out of business.

Editor's Note: The last paragraph of this article, regarding the political implications of H.R. 918, has been cut because the content essentially duplicates what is in the last paragraph on page 142 (article 27 in this volume).

30
HR 1096 Would Restrict Use of the Nation's Public Lands

REP. BRUCE F. VENTO (D-MN) introduced H.R. 1096 on February 21, 1991, in the U.S. House of Representatives. The bill's stated intent is "to authorize appropriations for programs, functions and activities of the Bureau of Land Management (BLM) for fiscal years 1992, 1993, 1994 and 1995; to improve the management of the public lands; and for other purposes." Vento states that the measure is a "fairly modest revision" of the Federal Land Policy and Management Act (FLPMA), which defines the Bureau's mission and duties.

In recent testimony on H.R. 1096 before the House National Parks Subcommittee, however, BLM Director Cy Jamison said the agency "strongly objects" to the enactment of the bill, which also reauthorizes BLM appropriations for the next four years. Jamison told the Subcommittee that the BLM wants the legislation amended because the agency believes many of the bill's provisions would change the character and thrust of the Federal Land Policy and Management Act from its policy of multiple use management to a policy of management by preferential consideration of selected resources. He said that H.R. 1096, unless amended, could prevent, or at least make difficult, the development of resources on public lands.

In early May, the House National Parks Subcommittee approved language in the bill that would make development of resources on public lands difficult, if not impossible. Section 8 of the bill amends FLPMA to require the Secretary of the Interior to "take any action necessary to prevent unnecessary degradation...to minimize adverse environmental impacts to (public) lands and their resources from use, occupancy or development...and to prevent impairment or derogation of the resources and values of conservation system units." Subcommittee Chairman Vento offered a judicial review amendment to the bill, which was adopted over the strong objections of Republican members, but no amendments to the provisions in Section 8 were offered during the bill's markup.

On May 22, the full House Interior Committee approved H.R. 1096 with provisions requiring BLM to emphasize the environmental values on the vast western public lands the agency administers still intact, over the heated objections of Republican representatives from the western states. The Westerners won a small battle, however, when a controversial amendment to drastically increase grazing fees on federal lands was withdrawn by its author, Rep. George "Buddy" Darden (D-GA). Darden withdrew his proposal to hike grazing fees due to lack of support from Western lawmakers of both parties.

However, Darden has said previously that he would try to take the proposal to increase grazing fees to the floor of the House when the BLM reauthorization bill is being considered. The measure could also be appended to the FY1992 budget appropriations authorization for the Interior Department, a route that was successfully used in the House last year.

During the markup session, the anger felt by Western Republicans was expressed in heated statements and efforts to delay approval of the bill. Rep. Don Young (R-AK) stated that H.R. 1096 amounted to "death to Westerners" and predicted that the legislation would die in the U.S. Senate.

"This is an open-ended hunting license for every environmental organization in the U.S.," Young emphatically stated. He was joined by Rep. Ron Marlenee (R-MT), who said, "This is litigation legislation at its worst."

Even after most Republicans stalked out of the markup session in an effort to jeopardize the quorum required for passing the measure, the bill passed easily on a voice vote by the Democrat majority.

Rep. Vento said the bill recognizes that the BLM has to put more emphasis on conservation. "You see the soil getting overgrazed, you see the dewatering of public lands, and these things become an open outrage," he said.

However, Director Jamison says the bill is an unacceptable "backdoor attempt" to change the federal land management act and the BLM's multiple-land-use mission. In the hopes that the legislation will be defeated, he has said, "We've lived without authorization for six years, and we can continue to do so."

Section 3 of H.R. 1096 reads (in part):

Section 103(a) of the Act (43 U.S.C. 1702[a]) is amended to read as follows: (a) The term 'areas of critical environmental concern' means areas within the public lands where special management attention (which may include restrictions on, or prohibition of, development) is required in order—(1) to protect important resources and values (including environmental, ecological, historic, cultural, scenic, fish and wildlife, and scientific resources and values) located on or likely to be affected by the use of public lands...(3) to protect or enhance the resources and values of a conservation system unit ...(b) Conservation Systems Unit—Section 103 of the Act (43 U.S.C. 1702) is amended by adding at the end thereof the following new subjection: (q) The term 'conservation system unit' means any unit of the National Park System, National Wildlife Refuge System, National Wild and Scenic Rivers System, National Trails System, National Wilderness Preservation System, or a National Conservation Area, , or National Forest Monument.

Section 5 of the bill states (in part):

(a)(2) Land use plans meeting the requirements of this Act shall be developed for all the public lands outside Alaska no later than January 1, 1997, and for all public lands no later than January 1, 1999...(b)(2) Section 202(c)(3) of the Act (43 U.S.C. 1712[c][3]) is amended to read as follows: (3) give priority to the designation and protection of areas of critical environmental concern and to identification, protection, and enhancement of the ecological, environmental, fish and wildlife, and other resources and values of riparian areas...(b)(3) Section 202(c)(5) of the Act (43 U.S.C. 1712[c][5]) is amended to read as follows: (5) consider present and potential uses including recreational and other non-consumptive uses of the public lands...

In other words, the bill changes the entire thrust of the Federal Land Policy and Management Act, placing the highest priority on environmental, ecological and recreational values. Natural resource development would have a much lower priority, and it would be subject to the land use planning process where it could easily be stringently restricted or prohibited. The BLM would be required to develop land use plans, rules and regulations based on the consideration of these "higher uses" of public lands, and thereby be forced into abandoning the multiple-use concept for natural resource development on the nation's public lands.

Fortunately, the chances of H.R. 1096 passing in the U.S. Senate do not appear to be likely. When a similar bill was easily passed in the House of Representatives in July 1989 it was strongly opposed by the federal administration, and it disappeared without even one hearing in the Senate.

For all intents and purposes, H.R. 1096 seems to be an environmentalist end-run attempt to further eliminate and restrict natural resource development on our public domain lands. Rep. Young's observation that this legislation would be "an open-ended hunting license for every environmental organization in the U.S." is right on target. So, miners and other natural resource users should start writing and calling their state's U.S. senators right away, and let them know that we are strongly opposed to this legislation.

31
HR 2614: Rep. DeFazio Attacks the 1872 Mining Law

HERE WE GO AGAIN! JUST WHEN miners were beginning to feel that the assault on America's mining community by the U.S. Congress had reached its peak, several more politicians decided to jump on the bandwagon. In an apparent attempt to appease the radical environmentalists (and most likely to also ensure their perpetual votes), a few more members of the Congress have decided to introduce their own version of mining law "reform."

On June 11, 1991, Rep. Peter DeFazio (D-OR) introduced H.R. 2614, the "Mining Law Reform Act of 1991
," in the U.S. House of Representatives. As is indicated by the bill's title, the legislative proposal is very similar to Sen. Bumpers' S. 433. This designer of H.R. 2614 obviously used S. 433 as a model, because the construction of many sections in both bills is nearly identical—including the exact same wording in many instances.

But the similarity stops there, Rep. DeFazio's bill is, in fact, much worse than Sen. Bumpers' bill. And it is much larger: There are 50 pages in H.R. 2614 vs. 36 pages in S. 433. While introducing his bill, Rep. DeFazio expressed his disdain for U.S. mining law in his remarks, a few of which are excerpted below:

> The 1872 Mining Law allows miners to obtain title to public lands for only $2.50 to $5 an acre. The law has been abused by speculators who obtained inexpensive development rights to valuable public lands. ...A 1989 study by the U.S. General Accounting Office (GAO) cited numerous claims across the West where land speculation was evident....
>
> The American taxpayer does not receive a fair return for the value of the minerals extracted from public lands. Under the 1872 Mining Law, companies remove precious minerals without paying a penny in royalties to the U.S. Government. The American taxpayer gets nothing in return.
>
> The 1872 Mining Law does not contain environmental protection or cleanup requirements. Reclamation standards and environmental controls depend on a flimsy patchwork of other Federal and State laws. Even where decent regulations exist, they are often impossible to enforce. A 1989 U.S. General Accounting Office study found more than 420,000 acres in 11 western states damaged by unreclaimed mining activity.
>
> My legislation eliminates the patenting of public lands and prohibits

individuals or corporations with more than 10 percent foreign ownership from mining on public lands. My bill also calls for a fair return to the public by setting the royalty rate at 12.5 percent, the same rate we apply to coal, oil, and natural gas extraction, and it creates a stringent environmental and reclamation permitting process. Finally, my bill establishes a hardrock mining impact assistance trust fund composed of 25 percent of the monies collected from the royalty. This fund will help rural communities mitigate the impacts associated with hardrock mining and help curb the traditional boom-bust cycle of mineral development in the West....

[T]alk is cheap. We've had this law on the books for almost 120 years. The General Mining Act may have served the national interest in 1872, but it's out of date today.

It is quite evident in his remarks that DeFazio based his homework on sensationalistic media reports and biased reports prepared by the GAO. It is also obvious that he didn't check out the *real costs* associated with the patenting process (normally above $1,000 per acre, and sometimes well over 10 times that amount). He also obviously discounts the fact that *miners pay every single tax* that other citizens and businesses pay, *in addition to some form of severance tax, net proceeds of mines tax, gross production tax, etc. that no one else is required to pay!* The congressman also could not have bothered to look up and read the overwhelming number of laws, rules and regulations that presently govern all mining activity, or checked into the total cost of the associated administrative charges, license fees, permit fees, regulatory fees or annual agency oversight assessments that are required by federal, state and local governments.

One of DeFazio's most outrageous statements was his remark that a 12.5% gross production royalty on hardrock minerals would provide a "fair return to the public." The hardrock mining industry is a high-cost, high-risk, capital-intensive business, and a 12.5% gross production royalty would equate to at least 30% of net income for most mining operations. When added to all other federal, state and local mining taxes and the federal income tax, this would increase the total tax bill at most mines to roughly 70% or more of their net income. Many marginal mines would actually be required to pay federal royalties when they are operating at a loss.

The royalty provision alone would force the closure of most of the nation's base-metal mines as well as a large number of industrial mineral producers and precious metals mines.

The stipulation that no mine operator could be more than 10%-owned by foreign corporations or stockholders would, of course, practically wipe out the U.S. mining industry. Americans are notorious for their unwillingness to invest in mining ventures, so the majority of the nation's mines are developed with a relatively high percentage of foreign capital. If this source of investment capital were to be, for all practical purposes, outlawed, then mineral exploration and development would come to an abrupt halt almost immediately.

As with Sen. Bumpers' S. 433, it appears that DeFazio's H.R. 2614 has been designed to effectively shut down mining in the United States—or, if it fails to achieve that objective, to impose huge fines and jail sentences on all miners who (heaven forbid) might occasionally make a mistake in their attempts to follow the overwhelming number of stringent new rules. It would be impossible to "fix" H.R. 2614—the bill is an absolute disaster—so miners can and should oppose the entire package.

Major Impact Analysis of H.R. 2614

1. As mentioned above, the participation of foreign investors, companies and individuals in mining activities on the public lands where 10% or more of the firm or stock is foreign-owned is disallowed. Because other provisions in the bill state the "effective date of this Act" is one year after the date of enactment of this act and a 3-year time period is allowed for conversion of existing mining claims to conform with the Act, it appears that any foreign firm or individual (or group of stockholders) with a 10% or more ownership in a U.S. mining operation on public lands would be forced into divesting themselves of that interest within 4 years of the date of enactment of this legislation.

2. Mining claim annual assessment/development work is replaced with a cash-only "Annual Holding Fee" of 12.5% federal royalty on gross production (whichever is higher), claimants must pay a $100 recording fee for each new claim; pay a holding fee of $5 per acre ($100 per 20-acre claim) per year in the first through the fifth year; pay $10 per acre ($200 per 20-acre claim) per year in years 6 through 10; pay $15 per acre during years 11 through 15; pay $20 per acre in year 16 and thereafter; pay additional user fees for admin-

istration of the Act, and pay an annual surface use fee of not less than $5 per acre for mill sites. This onerous burden of excessive cost requirements would ensure the elimination of all small-scale miners, prospectors, gold dredgers, small equipment manufacturers and peripheral businesses within 4 years following the date of enactment as well as the elimination of most of the U.S. mining industry shortly thereafter.

3. As with Title II of Rahall's H.R. 918 and Bumpers' S. 433, Title II of H.R. 2614 requires the imposition of excessively stringent environmental standards and reclamation requirements, all of which would apply to existing mining claims one year after the bill is enacted. There are again no time constraints in Title II, so it would automatically create virtually unlimited liability for all existing claimholders—in effect, Superfund II.

4. When it comes to provisions for civil and criminal penalties, DeFazio is truly astounding. Under Section 207, Enforcement, para. (d), it reads:

> Civil Penalties. —If any person fails to comply with any provisions of this Act, or any provision of a plan of operations...or any regulation or order issued under this Act...such person shall be liable for a civil penalty of not more than *$5,000 for each day* of the continuance of such failure.

Under Sec.207(e) is stated, in part:

> Criminal Penalties—any person who knowingly and willfully: (1) violates any provision of this Act, any provision of a plan of operations..., or any regulation or order issued under the authority of this Act... (2) makes any false statement, representation or certification in any application, record, report or other document filed or required to be maintained.... Or (3) falsifies, tampers with or renders inaccurate any monitoring device or method of record...shall, upon conviction, be punished by a fine of not more than *$100,000,* or by imprisonment for not more than 5 years, *or both. Each day* that a violation under clause (1) of this subsection continues, or each day that any monitoring device or data recorder remains inoperative or inaccurate because of any activity described in clause (3) of this subsection, *shall constitute a separate violation.*

Sec.207(f) states that all corporate or company officers or agents involved in a prosecution under (a) above may be liable for the same fines or imprisonment, or both, outlined above.

Sec.207(g) states that all of the remedies and penalties prescribed in this Act are concurrent and cumulative, and the exercise of one shall not preclude the exercise of the others. It also states that these penalties *"shall be in addition to any other remedies and penalties afforded by any other law or regulation."*

It is interesting to note that the bill's civil and criminal penalties could, in many cases, exceed the civil and criminal penalties allowed by law in the prosecution of hardened criminals.

5. As in the Rahall and Bumpers bills, Section 208 of Title II provides for "Citizen Suits" that specifically authorizes "any person having an interest *which is or may be* adversely affected" to file suit in a district court against the United States "or against any other person who is alleged to be in violation of any rule, regulation, order or permit issued pursuant to this title." This provision opens up any exploration or development project to perpetual litigation in the courts at any time and for any reason.

As Rep. Don Young (D-AK) recently said in reference to a different legislative proposal (H.R. 1096): "This is an open-ended hunting license for every environmental organization in the U.S." In fact, no sane person or company would invest time and money in a mining project with this type of unlimited risk—you could be shut down for just about any reason by anti-mining radicals, and hauled into court at any time.

Summary

The above analysis of major impacts of this legislative proposal only covers a few of the most notable provisions in the bill. As noted at the beginning of the article, H.R. 2614 was modeled after Sen. Bumpers' S. 433, so a major portion of the bill contains similar material. Both of these bills would sound the death knell for the nation's small-scale miners and signal the eventual end of mineral exploration and development in the United States. One very important point should be emphasized here: these legislative proposals illustrate very clearly the intention of certain members of Congress, and their extremist supporters, to permanently shut down all mining activity in the U.S.

32
Rep. Les AuCoin Launches Assault on Gold Miners

IN FEBRUARY OF THIS YEAR, Rep. Les AuCoin (D-OR) launched a vicious assault on the nation's gold miners with his announcement of the imminent introduction of "The Cyanide Mining Impact Relief Act of 1992" (H.R. 4298). A press release and a "Dear Colleague" letter from AuCoin, orchestrated behind the scenes to coincide simultaneously with rabid press releases by Defenders of Wildlife, the Sierra Club and the Mineral Policy Center, were carefully crafted to scare the hell out of the general public and members of Congress by implying that gold miners are creating a massive cyanide-laced environmental disaster throughout the United States.

For example, Rep. AuCoin's news release states, in part:

> A 50-cent per pound tax on sodium cyanide will curb the toxic and environmentally destructive practice of gold mining with cyanide by making it much more expensive, Rep. Les AuCoin (D-OR) and representatives of Defenders of Wildlife, the Sierra Club and the Mineral Policy Center said today.

The congressman's news release goes on to say:

> In the heap leaching process, mountains are demolished with explosives, then the rubble is sprinkled with cyanide solution, creating *huge toxic pools* covering several acres.

AuCoin adds in his Summary of Legislative Initiative:

> One significant problem is that thousands of waterfowl and other wildlife are killed each year because they are attracted to the cyanide ponds thinking that it is water. ...The purpose of the tax will be to provide a disincentive for using this nasty stuff and generate revenue which can be used to clean up the damage and research less environmentally harmful ways of extracting gold.

Rep. AuCoin's "Dear Colleague" letter (titled "Sodium Cyanide Mining: The Drift Nets of Our Public Lands") states:

> What this means is that in some operations entire mountains are moved to extract a few ounces of gold. This is not environmental impact, this is environmental removal... In addition, the cyanide solution which has been sprayed over mounds of ore is collected in large, open settling ponds. To airborne waterfowl they look like lakes. Cyanide-

laced holding ponds serve as a fatal attraction to wildlife and waterfowl. ...If these large, primarily foreign-owned mining corporations are going to have this kind of impact on our public resources, I think they should pay for it and I think they should clean it up...

Now, this is what Rep. AuCoin *did not say:*

1. The problems that his "cyanide tax" bill are purported to address *do not exist!* Any problems with wildlife mortalities and potential environmental contamination resulting from the use of cyanide in mining operations have already been addressed and eliminated. Cyanide use is presently under stringent regulation by federal and state agencies, miners already are *required* to completely clean up and reclaim sites where the chemical is used, huge bonds are already required to ensure that remediation takes place, the entire cost is currently being borne by miners, and *all leach ponds are inspected by regulatory agencies on a regular basis.*

2. The detoxification and reclamation requirements included in AuCoin's bill have been in effect for years. So, any revenues collected by his tax on cyanide would be used for something else.

3. AuCoin's $.50 per pound tax would *double* the cost of sodium cyanide to miners, effectively raising the cutoff grade of lower-grade ores to the point where they could not be mined at a profit. This would force marginal producers to shut down, thereby reducing economic activity, decreasing tax revenues, and creating higher unemployment—for no reason at all.

Again from his news release, AuCoin states: "We know what cyanide does to people: it kills them. And it does that to wildlife, too. But we don't know about the long-term impacts of cyanide on the environment, and we can't afford to take the chance while the mining companies get off scot free."

Here are the facts:

1. Cyanide has been used in mining for about a century and there are no reported cases of human fatalities from cyanide concentrations used by the mining industry.

2. Cyanide solutions can kill wildlife and there were problems with this in the past, but, according to wildlife officials, wildlife mortalities from cyanide use are negligible at the present time.

3. *We do know* about the long-term impacts of cyanide on the environment—*there aren't any,* because it breaks down rapidly upon expo-

sure to air and forms harmless compounds; and, finally,

4. If anyone is getting off "scot free," it isn't miners—they are bonded up to $10,000 per acre for sites where cyanide is used, they pay for all necessary neutralization, and they pay high costs for permitting any facility where cyanide is used.

Is a picture beginning to form here? A congressman is deliberately providing misinformation (that's the polite, PC term) to the American people and members of the U.S. Congress, based solely upon outright lies from the environmentalist groups (we'll get to this in a moment). Why is he doing this? Rep. AuCoin and radical extremists are attempting to create another "surrogate issue," whereby they can intentionally frighten the general public and members of Congress into doing their dirty work for them. That's right, folks, this is another blatant attempt to completely shut down all mining activity in the U.S.—permanently.

Now for a few select phrases from the news releases provided by environmentalists to the media in conjunction with AuCoin's missiles: The statement by James Dougherty, vice president of Defenders of Wildlife, includes the following: "...Irresponsible mining has left behind a trail of death... While the mining industry's use of cyanide is currently not regulated, the tax imposed by this [AuCoin's] bill will help make industry less careless with the use of this lethal chemical."

Where does Dougherty come from, some other planet? His outright lie to the media, general public and Congress that cyanide "is currently not regulated," coupled with off-the-wall references to "irresponsible mining," "trail of death," and "lethal chemical," is obviously designed to elicit outrage and horror from whoever happens to read this tripe. He then goes on to applaud Rep. AuCoin "for shedding some much-needed light on this critical environmental issue." In other words, he is attempting to convey the idea that this devastation of wildlife and the environment by miners is finally being recognized and addressed. The fact that the purported problem doesn't even exist shows how far these self-appointed saviors of mankind are willing to go to achieve their personal objectives.

In the news release put out by the Mineral Policy Center on the same day, MPC President Phil M. Hocker states: "Cyanide is the gold mining industry's dark side. Channeled and properly controlled, it can be beneficial." But, Hocker added:

> Hundreds of millions of pounds of this extremely poisonous chemical are being used across the West to take free gold from the public lands.... Gold on public lands may be free, but it's not cheap. Someone will have to pay billions to clean up the aftermath of today's gold boom. We need Representative AuCoin's bill, and the 'Mine Remediation Trust Fund' it creates, to make sure those who take the gold pay to clean up the mess.

Hocker is a little more careful in his avoidance of outright lies. He instead, by implication, leads the reader to believe that cyanide use in mining operations is not regulated, that it is poisoning the environment, and that most probably the taxpayer will have to pay the "billions to clean up the aftermath of today's gold boom." Clever, isn't he? Intentionally avoiding the truth, he utilizes carefully crafted phrases to mislead, misinform and dupe the public and Congress. Of course, we already know where Hocker is coming from (despite what he tries to con the public into believing): he just wants all mining to stop.

The rest of the Mineral Policy Center (wonder why they don't call it the Mineral Prohibition Center and come out of the closet?) statement includes the usual garbage, including:

> ...One teaspoonful of weak solution of sodium cyanide can be fatal to humans. The widespread use of cheap cyanide has made it profitable to strip-mine large quantities of gold ore with as little as a fiftieth of an ounce of gold per ton. This strip-mining for gold has brought large-scale disruption to many Western landscapes. ...Over twelve billion dollars' worth of gold will be taken from federal lands in Nevada in the next five years [not if you can help it, Phil], but under the 1872 Mining Law not a penny will be paid to the U.S. Treasury for the gold. ...The metals mining industry generates twice as much solid waste each year as all other U.S. industry and municipal landfills combined, but it is exempt from many national environmental laws...

I wonder if Hocker ever gets tired of repeating this constant litany of B.S. He *knows* that it will cost miners well over half the estimated $12 billion just to produce the gold, and he *knows* that a healthy chunk of any profits will be heavily taxed by the U.S. Government and the states. By the way, Hocker's definition of mining's "solid waste" is: If you break a rock in half it becomes waste, if you move a shovelful of dirt it becomes waste, and *if miners touch anything, it becomes solid waste.* This means, of course, that by the same definition all cities, roads, airports, farms, lawns, gardens, etc. are also composed of solid waste.

Now for some excerpts from the Sierra Club news release, provided by David Gardiner, the organization's legislative director:

> ...Today on our Western public lands, modern mining companies bulldoze and blast, using high-tech explosives, giant earthmovers—and a deadly chemical—to extract gold. Cyanide is used by modern operators to separate gold from low-grade ores in mines in Oregon, California, South Dakota, Nevada, Utah, South Carolina and other states. There are more questions than answers about the long-term effects on our groundwater; *the chemical is a toxic time-bomb that may not detonate for years....*
>
> There's no mistaking the effects on our wildlife. The shining water of a tailings pond beckons to waterfowl as they migrate across the arid land. But when the birds land, they find the glittering water is *laced with poison.* ...In short, industry gets the gold, and the public gets the shaft. Congressman AuCoin's bill is no substitute for the complete replacement of our current mining law, a statute written in a frontier era.
>
> ... Congressman AuCoin's legislation is yet another sign that Congress is ready to enact new legislation to fully protect our wildlife, our public lands, and our public health.

This guy is the craftiest one of the lot. Everything here is implied, and Gardiner doesn't actually *say* that anything bad is currently happening. But, the threat is definitely there in the words "deadly chemical," "toxic time-bomb," "laced with poison," and the final implied need for "new legislation to fully protect our wildlife, our public lands, and our public health." Pretty clever, eh? He scares the hell out of everybody without actually stating that there is definitely a current identifiable something to be afraid of.

As mentioned earlier, it becomes obvious that this entire furor over cyanide was designed to be another surrogate issue, with the ultimate aim of pressuring the American people and Congress into doing something stupid. Over the past decade or so, the leaders of the major environmental groups in the U.S. have publicly stated that they only want to ensure adequate protection of our environment, but they lately have dropped the façade and have gone all out in their efforts to destroy our nation's natural resource industries. This effort is to be utilized as another weapon in their arsenal—with the objective of blowing us out of the water before the general public and Congress can wake up and realize that they've been conned.

The battle never ends.

33
HR 918: Bill Would Stop Mining on Our Public Lands

REP. NICK J. RAHALL (D-WV) in June 1992 launched his expected revision of H.R. 918 in the form of an "Amendment in the Nature of a Substitute to H.R. 918." The first sentence following the title reads: "Strike all after the enacting clause and insert."

In other words, the long-anticipated revision to the original bill is a complete replacement, except for the bill number. As a result, the new "substitute" bill was lifted out of Rahall's subcommittee and marked up in the full House Committee on Interior and Insular Affairs on June 24—without any hearings held on the bill's contents. For that matter, it was placed on such a fast track that the legislation was pushed to the floor of the House of Representatives before many people could even obtain a copy of the bill. This process was obviously an end-run on the nation's minerals producers and their supporters in Congress, which was carefully orchestrated behind the scenes by Rep. George Miller, chairman of the Committee on Interior and Insular Affairs, and America's major environmental groups.

This is substantiated by the fact that the Mineral Policy (Police?) Center, mining's acknowledged enemy, issued a news release calling for environmentalist support for the legislation, including a complete description of the contents of the bill, on the same day the substitute bill was launched—and well before anyone else had any possible access to a copy of the bill. And it is no wonder that MPC President Phil Hocker fully supports the bill: H.R. 918's substitute amounts to an open-ended hunting license on America's miners for every anti-development, anti-everything radical extremist in the U.S. The Citizen Suit provision of Title II fully authorizes anyone, for any reason, at any time, to file suit against any individual miner or mining activity in the nation with full support and encouragement from our own federal government! And that's not all.

A careful examination of this "substitute H.R. 918" reveals that it contained just about every single demand made by radical environmentalists and practically none of the input provided by miners and the minerals industries. The original Title II in H.R. 918 last year has been expanded to include much more comprehensive coverage of the nastiest anti-mining provisions which were the primary focus of opposition by miners last year. In fact, the provisions of Title II, if enacted, would ensure the elimination of all mining activity on the nation's public lands for all time. This is obviously the express intent of Rep. Rahall, Rep. Miller, their supporters in Congress, and most of the major environmental groups.

This analysis is supported by a letter from Department of Agriculture Secretary Edward Madigan to Rep. Miller, which was delivered prior to the June 24 markup in the House Interior Committee. Mr. Madigan's letter says, in part:

> The Administration [is] opposed to H.R.918 as introduced at the hearing before the Subcommittee on Mining and Natural Resources (chaired by Rep. Rahall) on June 18, 1991, because the provisions of the bill affect the basic tenets of the mining law of access, self-initiation, and security of tenure. As we stated at that time, we believe that these concepts must be maintained in any mining legislation.
>
> Our serious concerns about the bill as introduced are part of the record, but we must add our equally serious concerns about the new and revised provisions of the amendment in nature of a substitute. Of particular concern to us are the provisions in Title II: Environmental Considerations of Minerals Exploration and Development. We object to the new provisions dealing with land use planning which require unsuitability reviews for mineral activities on all lands; the provisions for inspection and enforcement, especially those dealing with public participation, administrative hearings, and citizen suits; and the surface management provisions requiring exhaustive analysis and processes for permit application and inflexible reclamation standards.... We believe the cost of administering new procedures placed on the Federal government and the level of uncertainty placed on the mining industry from the new procedures and restrictions in this amendment will preclude operations on public lands....
>
> In summary, the Department of Agriculture has grave objections to the provisions of Title II of the amendment in nature of a substitute and objects to the operation of the amendment itself, because it affects the basic tenets of the mining law.

Furthermore, Secretary of the Interior Manual Lujan, Jr., sent a letter to House Interior Committee Chairman Miller on June 23 that included similar objections. Secretary Lujan's letter states, in part:

> This is to provide you with our views on the Rahall amendment in the nature of a substitute for H.R.918, the 'Mineral Exploration and Devel-

opment Act of 1991.' *The proposed amendment would make both substantive and technical changes to H.R.918* ...We strongly oppose the Rahall substitute to H.R.918. *If presented to the President as currently drafted, we would recommend that he veto the bill for the reasons discussed below.*

Please note that the House Interior and Insular Affairs Committee had received *strong opposition* to the H.R. 918 substitute from both the Department of Agriculture (U.S. Forest Service) and the Department of the Interior *prior to* the scheduled markup date. They had also received faxes and telephone calls from hundreds, if not thousands, of Westerners, including Gov. Bob Miller of Nevada.

So, what did the committee do? They passed the measure by a vote of 26 to 19, thus bowing to the supremacy of environmentalism in U.S. legislative policy. No small wonder, then, that the nation's most radical extremist elements are heaping praise upon those who obeyed their commands.

However, this is still not all of the story. During the markup of the bill, the committee also approved an amendment proposed by Rep. Peter DeFazio (D-OR) to deal the final death blow to the American mining industry. DeFazio's "Amendment to the Amendment in the Nature of a Substitute to H.R. 918" contains the following provisions:

> Sec. 410. ROYALTY: (a) RESERVATION OF ROYALTY. —Production of locatable minerals (including associated minerals) from any mining claim located under this Act, or mineral concentrates derived from locatable minerals produced from any mining claim located under this Act, as the case may be, shall be subject to a royalty of not less than 8 percent of the gross income from the production of such locatable minerals or concentrates, as the case may be.

Rep. DeFazio's amendment also states: (1) royalty payments are to be made no later than 30 days after the end of the month in which the mineral is produced in its first marketable condition; (2) stipulates reporting requirements including quarterly reports, records, documents and other data; (3) all persons holding claims are subject to financial audits; (4) 50% of royalties go to hardrock reclamation fund, 25% to the state in which the minerals are produced, and 25% to the U.S. Treasury; (5) claimholders are subject to fines and penalties (plus everything else under Sec. 202) if they are found in noncompliance with any requirements of Sec. 410, and (6) provision for more regulation and reports dealing with royalties.

This means that *all* miners are subject to the federal 8% royalty, including gold dredgers and panners, *whether they sell the mineral produced* or not, as soon as the mineral is produced in its first marketable form. This also means that *anyone* producing minerals of marketable value, whether sold or not, is required to file the reports, keep the records, be subjected to audits, and be fined and penalized for non-compliance. Each and every miner is also subject to investigation (and charges) by the newly appointed "enviropolice," which includes just about every rabid anti-everything radical in America.

The "new and improved" H.R. 918 also contains greatly expanded environmental and reclamation requirements and standards, some of which could be both technically and economically impossible to meet (especially since they are to be interpreted by federal bureaucrats and environmentalists). In effect, the combination of these nebulous requirements and standards creates an open-ended, indefinite timeframe for the (theoretical) final approval of any mining permit or application. *A miner could conceivably have to wait for up to 7-10 years before being allowed to initiate even a relatively small exploration project,* let alone the remote possibility that he may ever be allowed to actually open a mining operation. This is assuming, of course, that any person in their right mind would even consider entering into this costly, nerve-wracking and time-consuming process.

It is especially significant that the extended time delays and associated costs involved in the bill's new "permit application" and "reclamation plan" processes apply equally to large mining operations and to more routine matters, such as minor road construction, exploration trenching, or small exploration or mining projects (even suction dredging!), despite the fact that minor disturbances do not represent a potential for significant environmental impacts. These provisions in the new H.R. 918 alone provide an opportunity for anti-mining activists to preclude or eliminate any mining activity that they find offensive, regardless of merit or location, at their pleasure. In addition, they are given express permission to sue the pants off the individual miner whenever any actual mining activity takes place.

Get the picture here? By holding the ultimate authority over mining, radical extremists can prevent new mining from taking place, eliminate any existing mining activity, and place a tremendous economic liability on any miner or mining operation almost at will.

It is also now quite obvious that the numerous hearings held on the original H.R. 918 (and its predecessor, H.R. 3866) were a complete farce. Rep. Rahall and his supporters never intended to be fair with the hundreds of miners and mining industry representatives who provided thousands of pages of factual testimony on his so-called mining law "reform" proposals. In the final result, we can now see that this legislative disaster was prepared almost entirely under the direction of environmental radicals. Their outright lies, their deceit, their disinformation, and their theatrical bursts of emotional outrage are readily apparent in this "Amendment in the Nature of a Substitute to H.R. 918." If this incredibly biased bill is enacted by Congress, America's miners are slated for permanent unemployment in their chosen occupation unless they move to other countries.

How any member of Congress can even consider legislating away one of our nation's most productive industries, let alone intentionally destroying the livelihoods of thousands of American citizens, is difficult to imagine. This type of action is much more typical of the now-defunct Stalinist-Leninist Soviet Union or Hitler's Nazi Germany. Until now, it seemed almost inconceivable that such a thing could happen in the United States of America. Unfortunately, however, this type of legislative insanity is now appearing with alarming regularity in the U.S. Congress.

34
Bumpers' 'Amended' S. 433 Is a Trojan Horse

IMMEDIATELY FOLLOWING THE House Interior Committee's railroad job on Rep. Nick Rahall's "substitute" H.R. 918, Sen. Dale Bumpers (D-AR) introduced his version of an amendment/substitute to S. 433 in the U.S. Senate—whereupon the legislation was placed on a fast track for markup on July 1. On the surface, it then appeared that the "good guy" vs. "bad guy" roles between Sen. Bumpers and Rep. Rahall had been reversed.

Sen. Bumpers' "Amendment to S. 433" virtually gutted the original version by removing just about every objectionable provision in the bill and, as with Rahall's "substitute," replaced the full contents of S. 433 by inserting after the title: "Strike all after the enacting clause and insert in lieu thereof the following:... ." Procedurally, of course, this placed the bill on a fast track by retaining the original bill number and maintaining that the legislation had already been subjected to public hearings.

Guess what? To the uninitiated it would now appear that Bumpers has given up on his quest to impose all kinds of nasty controls on America's minerals industries and has actually decided to be reasonable about mining law "reform." Among the many problem provisions in the original version of S. 433, Sen. Bumpers' amendment deleted any reference to the following:

1. There is now absolutely no language in the replacement bill that would repeal, abolish or replace the 1872 Mining Law, as amended. In fact, the amended version refers several times to existing sections of the law as remaining intact and makes only relatively minor amendments to a few other sections of the law. There is no provision in the amended version that would take precedence over the current U.S. mining law, excepting the relatively minor tinkering involved in certain substitutions, small revisions, and some fairly innocuous additions. Although a few of these changes are objectionable, the amendment leaves the basic mining law intact.

2. Title II in the original bill (which used to be very similar to the Title II in H.R. 918 last year) has been eliminated in its entirety. In other words, S. 433, as amended, now contains no provisions for land

use, planning restrictions on mining activity, no provisions for civil or criminal penalties, no stringent environmental requirements, and no allowance for "citizen suits." In fact, all of the horror stories in the original bill have disappeared and, as if by magic, have reappeared in greatly expanded form in Rahall's "substitute H.R. 918."

3. The original section requiring the reclamation of abandoned mines, and the stipulation that today's miners should pay for it, has disappeared.

4. The rather unworkable requirement for the change in mining claim size, re-recording claims under the "reformed" laws, the attendant increases in fees, and penalties for infractions thereof, have also disappeared.

Now, here is a brief summary of what the amended version of S. 433 does contain:

1. *Annual Claim Holding Fee:* The bill substitutes an annual holding fee in lieu of the current annual labor requirement, in the following amounts: (1) $5 per acre in years 1 through 5, (2) $10 per acre in years 6-10, (3) $15 per acre in years 11-15, and (4) $20 per acre in each year thereafter.

2. *Claim Terms:* Mineral production in paying quantities to begin within 15 years of the claim is null and void, with a 5-year extension if claimant shows bona fide efforts to produce minerals in paying quantities. Provides for additional 15-year extension if production occurs for two consecutive years within the initial 15-year period or an extension thereof.

3. *Reclamation:* Requirements consist of: minimization of adverse environmental impacts, a plan of operations for activities causing more than a minimal disturbance (plan cannot be disapproved unless requirements cannot be met), and a fairly reasonable "shopping list" of reclamation requirements that must be included, and complied with, in a reclamation plan. Most notable here is the fact that backfilling of pits is not required.

4. *Bonding or Financial Guarantees:* A bond or other financial guarantee is required in amounts of not less than $200 or more than $2,500 per acre, but not less than the full estimated cost to complete reclamation.

Rahall's H.R. 918 "substitute" monstrosity consisted of 76 pages, but the entire bill now contains fewer than 10 full pages.

What is this? Is Sen. Bumpers actually backing off his earlier unreasonable and unrealistic position? No, far from it.

By making sure all of the nasty stuff was removed from his original bill, Sen. Bumpers hopes to enhance its quick and easy approval in the U.S. Senate. By normal procedures, this gutted version of S. 433 would then go to the House of Representatives for consideration on its own merit. However, these are not normal conditions.

You see, it is fully expected that Rep. Rahall's "substitute H.R. 918" will be passed by the House of Representatives and sent to the Senate for approval. Now, if both houses of Congress each pass their own version of legislation dealing with basically the same subject, then the two bills go to a Conference Committee where, theoretically, a mutually agreeable compromise is reached. Under normal circumstances, this is just a matter of thrashing out a few disagreements on relatively minor details. Not this time. Not by a long shot.

This year's mining law "reform" process in the U.S. Congress has been carefully crafted and planned behind the scenes by Sen. Bumpers, Rep. Rahall, Rep. George Miller and other anti-mining members of Congress with the close cooperation, coordination and direction of the nation's most radical environmental leaders. This is their big move to remove mining from the United States. *Now.* Remember the phrase "Mine Free by '93?"

The collaborators in this set-piece end run on the U.S. mining law decided to take all of the really nasty garbage out of the Senate, because they couldn't gain enough support to pass it, and placed it all in the new version of H.R. 918 in the House, where they could get enough strength to get it approved. By procedural maneuvering to move the "substitute" bill out of Rahall's subcommittee, they knew there was enough strength in the full House Interior and Insular Affairs Committee to add more onerous provisions (such as DeFazio's royalty amendment), mark up the new bill, and move it to a floor vote by the full House of Representatives without holding any hearings or discussion on the new provisions in the bill. They also knew that the majority of the House members had been completely brainwashed by environmentalist and media propaganda, so they were assured of getting the bill passed.

However, they also knew that the substitute H.R. 918 would be killed in the Senate if the bill moved through the normal committee process. So, they had Sen. Bumpers gut S. 433 of all of the most objectionable provisions by introducing his amendment as a substitute to S. 433. Again by keeping the same bill number, they hoped to move the bill out of the Mineral Resources Development and Production Subcommittee on procedural grounds (without holding any hearings) and move it to a vote of the full Senate Energy and Natural Resources Committee. From this point on, it should be a simple matter to get it passed by the full Senate.

Once Bumpers' gutted S. 433 is passed in the Senate and Rahall's new H.R. 918 substitute is passed in the House, both bills will have been removed from the normal committee process and placed in a House-Senate Conference Committee to work out the differences in the two bills. They would then be combined into one bill, and all of the nasty provisions in H.R. 918 would appear in the final product. Game over. All done. By circumventing the normal legislative process they would have achieved the ultimate objective of killing America's mining industry for good—or, failing that, handing over complete control of miners and mining to the radical environmental extremists who could most certainly finish the job.

What a masterful stroke for utopian environmentalism! And they could still pull it off it someone doesn't throw a monkey wrench into the process.

So, the scheduled markup of Bumpers' "Trojan horse" on July 1 should have (theoretically) gone without a hitch. However, some rational senators recognized this maneuver for exactly what it was, and they even called it a Trojan horse. As a result, no action was taken on Bumpers' amended S. 433 on July 1, and it was rescheduled for discussion on July 22. By not going along with (or being sucked in by) this farce, those senators who challenged this abortion of the political process have provided enough time (if utilized properly) to expose this legislative travesty to other members of Congress and the president.

Hopefully, this diabolical plot to destroy America's miners, the minerals industries and an entire productive sector of the U.S. economy will be killed in the U.S. Senate. Barring that, it is hoped that President Bush will emphatically veto the measure. If it comes to a presidential veto, it may be possible (although it is unlikely) for the House of Representatives to obtain the necessary two-thirds majority to override the veto. It seems almost impossible at this time that the necessary votes to override the veto could be achieved in the U.S. Senate.

Therefore, assuming that this legislative monstrosity remains alive for any appreciable period of time, it is absolutely imperative that all miners devote their time and effort to contacting their respective U.S. senators and President Bush with a clear (and loud) message that these bills (S. 433 and H.R. 918) are totally unacceptable.

35
Recent Action on Mining Laws in Congress

THE U.S. SENATE TOOK ACTION on August 5, 1992, on the FY1993 Department of the Interior appropriations bill, which contains several provisions dealing with amendments to the U.S. mining laws. A proposal by Sen. Harry Reid (D-NV) to amend the mineral patent provisions of the law was approved following a heated debate with Sen. Dale Bumpers (D-AR). Bumpers attempted to kill the Reid proposal, but was defeated on a vote of 52 to 44.

Sen. Reid's amendment, as approved, includes the following provisions:

> ...Mining Provisions: (1) Payment of fair market value—Any person receiving a patent pursuant to the Act commonly known as the Mining Law of 1872 (sections 2819 et seq. of the Revised Statutes) shall pay fair market value for the interest in the land owned by the United States exclusive of and without regard to the mineral deposits in the land.

Under paragraph (2), "Limitations," the amendment reads, in part:

> ...Any land patented after the date of enactment of this Act...shall be used only for mineral exploration, mineral development, mining, mineral processing, beneficiation, or uses reasonably incident to those uses, except with the approval of the Secretary.

Under paragraph (2)(B), "Reversion," it states, in part:

> ...Title to the land referred to in subparagraph (A) shall revert to the United States if the land is used for any unauthorized or unapproved use...

Under paragraph (2)(C), "Renouncing of Reversionary Interest," is stated, in part:

> ...If the Secretary finds that it would not be in the best interest of the United States to exercise the reversion for any reason...the Secretary may renounce the reversionary interest of the United States in the lands included in the patent... 8 years after an authorized use commences on the patented lands, the reversionary interest of the United States shall terminate.

In addition, under paragraph (3), "Reclamation," it states, in part:

> Any land patented after the date of enactment of this Act shall be subject to the mining reclamation law of the State in which the land is located. In the absence of applicable State mining reclamation law, the

land shall be subject to Federal mining reclamation law. Each patent shall recite that as a condition of the patent, the land patented shall be reclaimed to comply with Federal law or to comply with the mining reclamation law of the State in which the land is located.

Also approved by the Senate was an amendment proposed by Sen. Ted Stevens of Alaska to exempt certain small mining operations from the $100 mining claim holding fee. Congressional staffers report that the Stevens amendment consisted of the following: (1) Small mining operations which consist of 10 mining claims or less and which involve less than 10 acres in total land disturbance, and which are being actively mined, shall have the option of either paying the $100 fee per claim or performing annual assessment work of the same value. Unfortunately, no provision was made for a payment/work option on claims undergoing mineral exploration or evaluation.

The Senate Appropriations Committee also removed a proposed 33% hike in grazing fees and a mining patent moratorium provision. However, the House version of the FY1993 Interior Appropriations Act contains provisions for the increase in grazing fees, a mining patent moratorium, and the $100 mining claim holding fee. So, the differences between the House and Senate appropriations bills will be resolved in conference committee negotiations.

Both members of Congress and staffers report that the $100 mining claim fee is virtually a "sure thing," primarily because it now appears in both the House and Senate versions of the appropriations act as well as in the Federal Administrations FY1993 budget proposal. This means that Congress and the Administration have decided to let the nation's small-scale miners take the biggest financial hit, in spite of the overwhelming evidence that the $100 mining claim holding fee will practically devastate grassroots mineral exploration in the U.S. In effect, the appropriations legislation could now be called "The Small-scale Miner Unemployment Act of 1993."

As could be expected, the national news media showed its usual bias towards environmental extremism in most news items covering the Senate actions on mining law matters. In fact, they utilized the opportunity to restate the propaganda espoused by radical extremists, and most reports indicated their support for efforts to abolish the 1872 Mining Law, as amended.

On August 6, the Subcommittee of the House Agriculture Committee held a hearing on Title II of Rep. Rahall's H.R. 918. Interior and Agriculture representatives from the federal administration both provided testimony in opposition to H.R. 918, and it was reported that several Subcommittee members expressed their amazement when informed of the destructive nature of Title II. This is, of course, the primary reason behind Rep. Miller's and Rep. Rahall's efforts to get the bill passed without holding hearings. They knew that the real horror story contained in the "substitute bill" would be revealed if any hearings were held on the measure and, according to several reports from Washington, this is exactly what has happened.

It has also been reported that Sen. Bumpers has been unable to line up enough votes to get his S. 433 approved by the Senate Energy and Natural Resources Committee and has therefore directed his efforts towards making changes to the mining law through the Interior appropriations process. It is quite likely that he will be attempting to exert his influence on the House/Senate conference committee during its negotiations on the appropriations bills.

At this time (early August), it seems likely that Rahall's H.R. 918 and Bumpers' S. 433 will not be approved by Congress this year. It appears certain that the $100 mining claim holding fee will be approved, and it is likely that Sen. Reid's mining patent language will prevail.

36
Senator Bumpers Launches S.257 Anti-mining Missile

SENATOR DALE BUMPERS (D-AR) on Jan. 28, 1993, launched his multi-stage legislative and media anti-mining missiles in an attempted preemptive strike on America's minerals industries. Calling the 1872 Mining Law, as amended, "America's biggest ongoing scam" and "an outmoded relic" in his scathing press release, Bumpers initiated his annual blitz by introducing S. 257, entitled the "Mineral Exploration and Development Act of 1993," in the U.S. Senate. Co-sponsors of the bill included: Senators Tom Harkin (D-IA), Russell Feingold (D-WI), James Jeffords (R-VT), Herb Kohl (D-WI), Frank Lautenberg (D-NJ), Patrick Leahy (D-VT), Carl Levin (D-MI), Barbara Mikulski (D-MD), Claiborne Pell (D-RI), David Pryor (D-AR), Don Riegle (D-MI), and Paul Wellstone (D-MN). Because he was able to include one Republican with 12 Democrats backing the legislation (including himself), Bumpers felt justified in calling the proposal "bipartisan legislation," in spite of the fact that almost all of these senators represent the eastern half of the U.S.

The fact that 13 U.S. senators (or 13% of the U.S. Senate) have agreed to support this legislation would appear to indicate a significant amount of concern over the mining laws, right? *Wrong!* The 1872 Mining Law, as amended, provides for the exploration and development of mineral deposits on the nation's *public domain lands*. The total amount of public domain lands in the nine states represented by these senators is 2,550,852 acres, or about four-tenths of one percent (.0043) of America's total land area under the jurisdiction of the 1872 Mining Law, as amended. So, they seek to regulate mining activity on the 99.6% of the nation's public domain lands that lie outside of the boundaries of the states they represent (for that matter, three of these states have no public domain lands at all).

Please note that roughly 95% of our country's public domain lands are located in 11 western states and Alaska (where, as a consequence, nearly all of the mining activity under the mining laws actually occurs), and that not even one senator from these states has cosponsored this bill. Quite the opposite, in fact. Many western senators (both Democrat and Republican), in a *real* bipartisan effort, have vehemently opposed Bumpers' efforts to completely abolish the 1872 Mining Law, as amended, during the past four years. This is because the Westerners know that the real objective of Bumpers' so-called "mining law reform" is to completely shut down all mining on America's public lands, thereby destroying an industry which is vital to the economic health of the West.

In addition to a plethora of other unbelievable provisions in the bill that will be described below, S. 257 also imposes an 8% gross production royalty on all hardrock minerals. Actually, Section 410(a) specifies "... a royalty of *not less than 8 percent* of the gross income...." Considering the insatiable appetite of our federal government for "enhancing revenues" to fund even more insatiable social programs, it appears that this royalty provision might have become the prime mover of the legislation.

During the January 28 news conference he arranged to promote his legislative proposal, Bumpers said the election of former Arkansas Gov. Bill Clinton as president gives his bill a better chance for passage this year, even though Clinton didn't take a position on mining during the presidential campaign and Bumpers says he has not yet talked to him about it. Sure. The senior senator from Arkansas has been publicly attacking mining and the mining laws for years now, and he says he hasn't talked to the governor of Arkansas about it?

"But I know his values," Bumpers said to the media, "and I know this deficit is of such a critical importance that where you have a chance to pick up a hundred million dollars a year, we're going to do it. We're not going to continue to allow that kind of revenue to go uncollected."

Now we hear reports from several knowledgeable government officials that President Clinton's new federal budget contains an 8% gross production royalty on hardrock minerals and the resultant revenues projected therefrom are included in the administration's "economic stimulation" spending plan. Although this report cannot be absolutely confirmed at this time, Clinton is scheduled to deliver his new and improved federal budget to Congress sometime in March, so miners should know for sure at that time. If the royalty provision is, in fact, in the budget, wouldn't this be an amazing coincidence? Might there be some form of telepathic communication here?

Now for another amazing coincidence: Rep. Nick Rahall's H.R. 322 and Sen. Bumpers' S. 257 are virtually identical. Approximately 95% of both bills are word for word and title for title (even having the same section numbers and paragraph and subparagraph designation, excepting some errors Rahall made in Title III). The *only* major difference in the two legislative proposals' contents is found in Title I, Section 104, where Bumpers has eliminated the "diligence" work provisions.

This can be summarized as follows: Sen. Bumpers only wanted money to be paid to the government and could care less about actual exploration and development work on the ground (actually *develop mineral deposits?* What an outdated concept!). Rep. Rahall, on the other hand, allows diligent development expenditures to count toward the payment of the annual claim rental fees to the government, excepting a specified minimum annual rental fee of $2.50 per acre. Bumpers also adds some different language to Sec. 104 relating to relinquishment, declaring claims null and void, and some other provisions related to claim maintenance.

As in Rahall's predecessor bill, H.R. 918 (and subsequent "substitute"), both of the new bills call for annual rental fees as follows: $5 per acre in years 1-5; $10 per acre in 6-10; $15 per acre in 11-15; $20 per acre in 16-20, and $25 per acre for the 21st year and each year thereafter.

However, the annual mining claim rental fees and the 8% gross production royalty are both somewhat superfluous to most of America's miners, because it is extremely doubtful that anyone would be stupid enough to continue holding mining claims (let alone actually mine minerals) on the nation's public domain lands anyway. This is quite adequately assured by the provisions contained in both Bumpers' and Rahall's bills under Title II. A full explanation of the horror story contained in this title may be found in the *CMJ* article, "H.R. 322: Rep. Rahall's 'Mine Free By '93' Bill." This title (Title II) is identical in both bills.

Because S. 257 and H.R. 322 are almost identical, they could be quite easily merged together in a House/Senate Conference Committee (probably in less than an hour) if they are both passed. This is also positive proof that Sen. Bumpers and Rep. Rahall (as well as radical environmentalists and other anti-mining members of Congress) have been working very closely together in a coordinated effort to abolish the 1872 Mining Law, as amended, in 1993. If the 8% royalty is, in fact, included in the president's budget, it also shows even more behind-the-scenes planning and coordination by those who wish to eliminate mining in the United States.

Most importantly, if the 8% royalty is included in the federal budget, *it will be almost impossible to get it removed.* In this case, Sen. Bumpers is absolutely correct: the U.S. Congress is not about to let *any* "kind of revenue...go uncollected," even if it destroys an entire industry and an important sector of the U.S. economy. If any projected revenue has been included in the budget, Congress will spend it! So, as previously mentioned, the search for potential "revenue enhancement" (by any means that doesn't use the term "taxes") for increased federal spending may become the major factor driving the so-called mining law reform effort.

If anyone out there still thinks that our congressional leaders would not knowingly create such a tremendous negative impact on U.S. mining, remember what they did with the $100 mining claim fee last year. In spite of over three years of input by miners that graphically illustrated exactly what the fee would do to U.S. mineral exploration and its extremely negative impact on the small mining community, it was approved with very little debate or opposition. Knowing full well that the fee would not produce the revenues projected in the budget, Congress went ahead and approved the measure and spent the money. Mining claim activity has already dropped sharply, many major mining firms have made drastic cuts in their U.S. exploration budgets for 1993, grassroots exploration is practically nonexistent, and more mining companies are planning to move offshore—*and this is occurring even before the new claim fee has to be paid.* And, if there are still any members of Congress who actually expect this fee to produce any significant revenues, they are due for a rather rude awakening by late September or early October of this year.

The introduction of Sen. Bumpers' new and improved S. 257 also proves that his 1992 bill (the "substitute" S. 433) was truly the Trojan horse it appeared to be. Now he has dropped the façade of attempting reasonable reform to U.S. mining law and is back to showing his true colors. This is especially evident in the tripe that he is feeding the media. Consider the following from his January 28 news release:

> According to the latest figures, the federal government has virtually given away $94 billion worth of land just in the past six years. And every single year, $4 billion worth of minerals belonging to the American people is carted off and profitably sold—without a dime going to the U.S. Treasury. The whole process is so outrageous that it's almost unbelievable. It must be changed.

What planet does this guy live on, anyway? "...without a dime going to the U.S. Treasury?" He knows that mining pays *all* of the taxes paid by all other types of business and individuals, *in addition* to the state taxes and royalties miners pay on minerals production that no one else pays. And that's not counting the taxes paid by employees and peripheral businesses, or the additional revenues generated by economic activity created in the processing of mineral materials into finished goods (or the sale of same). In other words, this statement is absolutely untrue.

Sen. Bumpers also asserts, "Under the 1872 law, claimholders can obtain claims to virtually any federal land with mineral deposits." Come again? Several reports to Congress during the 1980s specifically point out the fact that well over 50% of the nation's federal lands are closed to any type of mineral development. There are presently a plethora of administrative land withdrawals where mining is excluded, and massive tracts of public domain lands have been withdrawn from operation of the 1872 Mining Law, as amended, by Congress.

And, as usual, Bumpers continues to say that miners can easily buy federal lands "for as little as $2.50 an acre."

It is quite obvious that Sen. Bumpers is trying to further enrage the American public by feeding them false information about mining, with the objective of gaining widespread support for approval of S. 257. It is also obvious that Bumpers has adopted the radical environmentalists' motto of "Mine-Free by '93."

37
HR 322: Rahall's "Mine Free by '93" Bill

IN JANUARY OF THIS YEAR, Rep. Nick J. Rahall II (D-WV) fired the first shot in the battle to eliminate mining on the nation's public lands by introducing H.R. 322, entitled the "Mineral Exploration Development Act of 1993," in the 103rd Congress. Co-sponsors of the bill included Reps. Richard Lehman, Bruce Vento and George Miller, the latter of whom is chairman of the new Committee on Natural Resources (previously the Committee on Interior and Insular Affairs).

In his introductory remarks, Rep. Rahall stated that the legislation is based upon the "Amendment in the Nature of a Substitute to H.R. 918," which he introduced last year and almost successfully railroaded through the House of Representatives without any public hearings on the measure. However, this monstrosity continues to grow. While the H.R. 918 "substitute" consisted of 76 pages, the new and improved H.R. 322 contained 105 pages.

As might be expected, over 50% of this legislative expansion (an additional 15 pages) occurred in Title II, "Environmental Considerations of Mineral Exploration and Development"—which means that over half of this so-called "mining law reform" measure (54 pages) is devoted to environmental issues.

Title II of H.R. 322, as written, is sufficient to shut down all mining on U.S. public lands within just a few years. Upon careful review of the provisions contained in this Title, it becomes readily apparent that it actually consists of: (1) an open-ended "hunting license" on miners for every radical extremist in America, (2) total control of mineral resource development by environmental groups, (3) virtually unlimited liabilities for all U.S. mineral producers, and (4) an extraordinary increase in time and costs that would negate the economic feasibility of developing almost all mineral deposits.

And that's not all, not by a long shot. For those who are stupid enough to attempt to produce minerals under this legislative insanity, there is also a provision for an 8% gross production royalty—which appears in Title IV, Section 410, "Royalty," as follows:

> (a) Reservation of Royalty.—Production of locatable minerals (including associated minerals) from any mining claim located or con-

verted under this Act, or mineral concentrates derived from locatable minerals produced from any mining claim located or converted under this Act, as the case may be, shall be subject to a royalty of not less than 8 percent of the gross income from the production of such locatable minerals or concentrates, as the case may be.

(b) Royalty Payments. —Royalty payments shall be made to the United States not later than 30 days after the end of the month in which the product is produced and placed in its first marketable condition, consistent with prevailing practices in the industry."

Section 410 also contains requirements for "quarterly reports, records, documents and other data....pertinent technical and financial data relating to the quantity, quality, and amount of all minerals extracted..." and a provision for audits of all claimholders.

Note that the royalty is set at "not less than 8%," which allows the Congress to increase the percentage at will, at any time. In addition, it takes at least half of any mine's total revenues for operating and capital costs associated with hardrock minerals production, *which means the minimum royalty rate would equal at least 16% of net income.* This would completely eliminate the average net profits margin after taxes for most U.S. mines. Keep in mind the fact that all U.S. mines currently pay some type of production, severance or net profits tax to the state in which each mine is located, *in addition to all of the taxes paid by other businesses.* With all profit incentive removed, even existing profitable mines would be forced to shut down, and no investor in their right mind would even consider financing a new mining project (actually, the provisions of Title II would effectively eliminate mine financing, even without the imposition of a gross production royalty).

Because the 8% gross production royalty is imposed on *all* minerals production, almost all small-scale mining operations and all marginal mines would, of course, be immediately wiped out. Many of these enterprises are currently operating at a loss or near the break-even point, which effectively means that the operator would be paying a federal royalty on money expended for operating or developing costs. Miners would also be paying the U.S. Treasury an 8% royalty on all taxes paid to the state in which the mine is located (double taxation). No other business in America is subjected to this type of treatment.

In his introductory remarks, Rep. Rahall said, in part:

> ...First, the bill recognizes that self-initiation and access to public domain lands open to the location of mining claims are important features of the Mining Law of 1872 that should be maintained. This is a mining claim bill, based on the principles of access to public domain lands and the right of self-initiation....
>
> This bill says to the prospective mining claimant that once a claim is properly located it is the exclusive possession of the locator for mineral prospecting and mining purposes so long as he is being diligent, pays the annual rental, and files an affidavit once a year.

Sure. While the bill is purported to provide security of title to mining claims, it also effectively eliminates security of title to mineral deposits by making all types of mining activity subject to the land use planning and review processes. According to sections 201, 204 and 205 of Title II, the actual mining of mineral deposits may be disallowed at the discretion of land management agencies and/or by objection from any member of the general public during the review of proposed mining plans of operations and the agency land use planning process.

Section 204, "Unsuitability Review," directs the Secretary of Interior and Secretary of Agriculture to "conduct a review of lands that are subject to this Act in order to determine whether there are any areas which are *unsuitable for all or certain types of mineral activities...*"

It then goes on to list six full pages of open-ended reasons for which lands may be found to be "unsuitable" for mining activities. Under the conditions listed in this section, any or all land in the United States could be found to be unsuitable for mining.

Under Section 201(b) of Title II,

> ...no person may engage in mineral activities that may cause a disturbance of surface resources unless such person has filed a plan of operations with, and received approval of such plan of operations, from the Secretary...

except for certain exploration activities that cause a negligible disturbance and do not involve:

> ...the use of mechanized earth-moving equipment, suction dredging, explosives, the use of motor vehicles in areas closed to off-road vehicles, the construction of roads, drill pads, or the use of toxic or hazardous materials.

This means that most small exploration projects and small-scale mining operations would be subjected to provisions of the National

Environmental Policy Act (the NEPA process), public hearings, expensive permits, long time delays, full bonding/surety, the land-use planning process, and possibly an Environmental Impact Statement (EIS). Only the very rich need apply, because this is a *very* expensive and time-consuming process.

In addition, under Section 201(c) is stated the following:

> Contents of Plans. —Each proposed plan of operations shall include a mining permit application and a reclamation plan together with such documentation as necessary to ensure compliance with applicable Federal and State environmental laws and regulations.

There is no distinction between exploration activities and mining operations nor, for that matter, between small and large projects.

To illustrate the overwhelming complexity of these requirements, consider trying to meet just a couple of the "Mining Permit Application Requirements" under Section 201(d) in a Plan of Operations filed for a small suction dredging operations, as follows:

> (7) A description of the quantity and quality of surface and ground water resources within and along the boundaries of, and adjacent to, the area subject to mineral activities, *based on 12 months of pre-disturbance monitoring*.
>
> (8) A description of the biological resources found in or adjacent to the area subject to mineral activities, including vegetation, fish and wildlife, riparian and wetland habitats.
>
> (9) A description of the monitoring systems to be used to detect and determine whether compliance has and is occurring consistent with the surface management requirements and to regulate the effects of mineral activities and reclamation on the site and surrounding environment, including but not limited to, groundwater, surface water, air and soils.

If this isn't enough to scare the pants off any prospective applicants for a mining permit, consider the provisions under Section 201 (1):

> Bonds. —(1) Before any plan of operations is approved pursuant to this Act, or any mineral activities are conducted pursuant to subsection (b)(2) (under a notice), the operation shall file with the Secretary financial assurance payable to the United States and conditional upon faithful performance of all requirements of this Act. The financial assurance shall be provided in the form of a surety bond, trust fund, cash or equivalent. The amount of the financial assurance shall be sufficient to assure the completion of reclamation satisfying the requirements of

this Act if the work had to be performed by the Secretary in the event of forfeiture, and the calculation shall take into account the maximum level of financial exposure which shall arise during the mineral activity including, but not limited to, provision for accident contingencies.

A careful review of the reclamation standards and requirements (Sec. 201(m)) contained in the legislation reveals that, for all practical purposes, they may be adequately summarized in the following statement: "Restore all mined land disturbances to their natural state." Considering the reclamation requirements alone, even without the unlimited liability provisions in the bill, *no commercial bonding firm would consider touching a mine reclamation bond with a ten-foot pole.* As a result, miners would be forced into posting their own bonds (no bond pool could conceivably accept the liabilities involved, either).

And then there's the real clincher. Under Section 202 "Inspection and Enforcement," subparagraph (e) "Citizen Suits," is stated, in part:

> (1) Except as provided under paragraph (2) [Note: This says no action may be commenced until 60 days after a plaintiff has given notice in writing of an alleged violation to the Secretary or to the person alleged to be in violation], any person having an interest which is or may be adversely affected may commence a civil action on his or her own behalf to compel compliance—(A) against the Secretary—(B) against any other person alleged to be in violation of any of the provisions of this Act or any regulation promulgated pursuant to this Act or terms and conditions of any plan of operation approved pursuant to this Act;...

This is the open-ended "hunting license" provision widely celebrated by the environmentalists. It specifically invites every anti-mining, anti-development, or anti-everything radical extremist in the U.S. to file suit against any individual miner or mining activity at any time, for any reason, with the full encouragement and support of our federal government. Get the obvious intent here? By being given the ultimate authority over all mining activity, radical extremists could now prevent any new mining from taking place, eliminate any existing mining activity, and impose a tremendous economic liability on any miner or mining operation almost at will.

It should now be quite obvious that this legislative disaster was prepared almost exclusively under the direction of environmental extremists. Their outright lies, deceit, and disinformation, and their theatrical bursts of emotional outrage during the numerous hearings held on the original H.R. 918 (and its predecessor, H.R. 3866), have been incorporated in the provisions of H.R. 322. If this incredibly biased bill is enacted by Congress, America's miners are slated for permanent unemployment—unless they move to other countries, as many U.S. businesses are doing.

38
Senator Craig's S.775 Clears Senate Committee

THE SENATE ENERGY AND NATURAL RESOURCES Committee on May 6, 1993, approved Sen. Larry Craig's S. 775 bill to reform the 1872 Mining Law, as amended. Acting within minutes and without any debate, the committee approved the mining industry-supported measure without making any changes. The bill will now move to the floor of the Senate for final approval, and it is expected that the bill will pass without debate or amendment.

However, Sen. J. Bennett Johnston, Chairman of the Energy and Natural Resources Committee, stressed that the committee's action was part of a strategy to push the mining issue through the Senate without consuming time and energy in battles over the final resolution of mining law reform. Under this strategy, substantial work on the bill will be conducted later this year by a small group of Senate and House negotiators in conference committee.

Sen. Dale Bumpers, sponsor of the onerous S. 257 mining law reform bill, concurred with the strategy, as did the Clinton Administration. This is because they felt that Bumpers' bill had little chance of approval with the expected strong opposition by Western senators. They are expecting the House of Representatives to pass Rep. Nick Rahall's stringent H.R. 322 and send it to the conference committee, where opponents of mining plan to produce a much more stringent overhaul of the nation's mining laws during the process of working out so-called "differences" between S. 775 and H.R. 322.

Although Sen. Craig accurately described his bill as accomplishing substantial mining law reform, he was immediately blasted by environmentalists who vehemently opposed any reasonable solutions to reform of the 1872 Mining Law, as amended. They, of course, fully support Bumpers' S. 257 and Rahall's H.R. 322 (which are virtually identical) because these bills would put an end to mining in the United States.

For example, Kathryn Hohmann, a Sierra Club director, called the Craig bill one "only a miner or his mother could love. Once again the miners get the gold and the American people get the shaft."

Sen. Craig and six other Republican cosponsors of S. 775, with support from several Western Democrats, are obviously attempting to accomplish a rational and realistic reform of U.S. mining laws that the mining industry can at least live with. However, radical and power-hungry environmentalists feel that they have the full support of the Clinton Administration and the majority of the U.S. Congress, so they are not about to consider a reasonable compromise on the mining law issue. They want to kill mining in the U.S. and will accept nothing less.

For example, during the Senate Energy and Natural Resources Committee hearing on royalties on Tuesday, May 4, Mineral Policy Center President Phil Hocker was the only person appearing before the panel that absolutely insisted on implementing a 12.5% gross production royalty on mining. When questioned about the royalties imposed on mining in other countries, he admitted that he didn't know but would look it up. One senator asked him directly how he could testify on royalties when he knew absolutely nothing about the subject.

Even Interior Secretary Bruce Babbitt said that the Clinton Administration is willing to go along with the 8% gross production royalty contained in Bumpers' S. 257 and Rahall's H.R. 322. This is a strong indication that Clinton's crew and mining's opponents are planning a behind-the-scenes strategy to load up any so-called mining law reform bill with an 8% gross production royalty and other onerous requirements in the House/Senate conference committee.

Unfortunately, this is beginning to resemble the Bumpers/Miller/Rahall plan last year. Sen. Bumpers introduced his watered-down "substitute" bill in an effort to obtain Senate approval of a Trojan horse measure that would meet Rep. Rahall's bill in a House/Senate conference committee, whereupon they planned to again "substitute" Rahall's "substitute bill" for the Senate-approved version. Since Bumpers agreed to the strategy on approving Sen. Craig's S. 775, it appears very likely that last year's plan to approve Rep. Rahall's bill in conference committee is still very much alive.

So while it may appear on the surface that the successful approval of S. 775 in the Senate provides an opportunity for reasonable reform of the mining law, the bill may actually be serving the same Trojan-horse purpose that Sen. Bumpers' bill was intended to serve in the last session of Congress. In any event, it is extremely unlikely that the final version of mining law "reform" coming out of the conference committee will be anywhere close to the current contents of Sen. Craig's bill.

Considering the current political situation, it is absolutely certain that an extremely stringent mining law reform bill coming out of the House/Senate conference committee would be immediately approved by the House of Representatives and signed into law by President Clinton. The question is, will the full Senate vote to approve it?

Editor's Note: According to information obtained in online research just prior to publication of this volume, the Clinton Administration's "mining law reform" efforts discussed in this article collapsed in 1994.

Afterword

With article titles in this volume such as "Miners Are a Threatened and Endangered Species" and "Extremists Mount Offensive in Environmental War," it's clear that Dave W. Parkhurst was not hesitant to call things as he saw them. As the content of Volume 3 evidences, what he saw was an honorable, productive industry—and one that is critical to the prosperity and even the survival of this nation—under relentless attack by radical environmentalist groups, anti-development factions, overzealous regulators, politicians seeking favor with these entities, and a biased news media.

The forces arrayed against U.S. mining and the natural resource industries in the 1980s and early 1990s (the time period during which Dave was chronicling the events, actions and conditions affecting and threatening those industries) were many and strident, and the voices in their defense seemed all too few and faint. Dave was one of those who did make his voice heard in the fight to help preserve the viability of America's metals and minerals industry, in particular, as well as the rights and opportunities that small miners and prospectors had enjoyed for well over a hundred years.

An avid outdoorsman, Dave thrived in Nature with its many geological wonders and challenges. With his lifelong history of being a responsible user of the public lands and his affinity for the natural environment and wildlife, he bore no resemblance to the characterization of miners by their opponents as oblivious or hostile to the need for environmental protection and conservation. Neither did his mining friends, partners, and professional associates.

The political, legislative and regulatory battles that Dave and other pro-mining, pro-fairness activists fought were focused primarily on preserving the 1872 Mining Law, as amended, and defending against regulatory excess. As a writer and lobbyist, Dave sometimes had to immerse himself in legalese and bureaucratese; this was likely his idea of hell at times, but as in everything he undertook he did "whatever it takes to get the job done." He never quit fighting for what he believed in, most especially the cause he had fought for passionately throughout his life: individual liberty and preservation of the rights guaranteed to Americans under the U.S. Constitution. He was an unfaltering warrior for that cause until the day he died.

About the Articles' Author

An in-depth profile of the author of the articles comprising the Dave W. Parkhurst Mining Writing Collection is presented in Volume 1 of the collection. More about Dave and his work as a mining writer, advocate and practitioner can be found in the Afterword sections of the four volumes in the collection and at **www.PineNutPress.com**.

Dave's lifelong love of prospecting and exploring for valuable minerals led to his entering the mining field full-time in early 1980 after many years spent working in other occupations, particularly telecommunications. He spent most of the remaining 13 years of his life writing about various aspects of mining and prospecting, as well as consulting for mining companies and lobbying in behalf of small-scale miners and prospectors. His first love, though, was being out in the hills prospecting for gold and then developing his prospects.

On a golden fall day in September 1993, nine days after his 55th birthday, Dave was finishing up some work in his home office in Minden, Nevada. He was looking forward to the upcoming weekend when he could go out to the mine he co-owned with several partners to continue the gold exploration and recovery operation he and one of the partners were working on. Suddenly he collapsed, having suffered a massive heart attack; he died after paramedics' efforts to revive him were unsuccessful.

Tributes to Dave recognizing his contributions to mining and the small-mining community in the American West have been reprinted in Volumes 1 and Volume 4 of the Dave W. Parkhurst Mining Writing Collection.

APPENDIX A

GOLD PLACERS AND MINERAL DEPOSITS: Their Formation, Deposition, and Characteristics is the first of the four volumes that make up the Dave W. Parkhurst Mining Writing Collection. It encompasses four main topics:
- How, where and why mineral deposits form.
- The formation and characteristics of placer deposits.
- The nature, characteristics and concentration of gold and where to look for gold.
- Prospecting for valuable metals and minerals other than gold.

THE BASICS OF GOING FOR THE GOLD: From Prospecting and Exploration to Small-scale Mining Project, the second volume of the collection, consists of articles on the actual mechanics and processes of prospecting and mining, primarily for gold. It is divided into six main topics:
- Searching for gold.
- Preparing to explore and prospect.
- Prospecting, sampling, and evaluation.
- Recovering the gold and other values.
- Prospecting and mining miscellany.
- Case studies and small mining projects.

A CRITICAL INDUSTRY UNDER ATTACK: The Struggle to Preserve Metals and Minerals Mining Viability in the U.S. is Volume 3 in the set. The three main parts of the book concern these topics:
- The issues, challenges and threats affecting U.S. mining in the 1980s and early 1990s.
- Miners and mining vs. anti-mining extremism.
- Defense of the 1872 Mining Law as amended.

FIGHTING THE GOOD FIGHT: Mining's Battle for Survival in the American West is the last of the four volumes in the Dave W. Parkhurst Mining Writing Collection. It covers five main topics:
- The U.S. government vs. mining.
- Miners and the U.S. Forest Service.

- Implementation of the mining claim holding fee.
- Nevada miners vs. the politicians and anti-mining factions.
- General articles related to mining in the Silver State.

The four volumes in the Dave W. Parkhurst Mining Writing Collection are available at Amazon.com and elsewhere. For additional information about the books in this collection or about Dave Parkhurst and his work, please visit **www.PineNutPress.com**.

APPENDIX B

YEAR OF PUBLICATION
IN THE *CMJ*

Article No.	Article Title in Alphabetical Order	Pub. Year
7	1987 Was a Good Year (?)	1987
1	America Needs a Strong Minerals Industry	1990
29	Analysis of H.R. 918, Rep. Rahall's New Mining Law Bill, An	1991
28	Analysis of S. 433, Senator Bumpers' Mining Law Reform Bill, An	1991
19	Anti-mining Propaganda: Half-truths and Lies	1992
8	BLM Reclamation Bonding and the Small Miner	1988
34	Bumpers' "Amended" S. 433 Is a Trojan Horse	1992
6	Environmental "Protectionism"	1987
14	Environmental Activists Organize to Eliminate Mining Laws	1989
13	Extremists Mount Offensive in Environmental War	1988
12	FederalSpeak Dictionary	1993
30	H.R. 1096 Would Restrict Use of the Nation's Public Lands	1991
31	H.R. 2614: Rep. DeFazio Attacks the 1872 Mining Law	1991
37	H.R. 322: Rep. Rahall's "Mine Free By '93" Bill	1993
26	H.R. 3866: Rep. Rahall's "Mineral Exploration & Development Act of 1990"	1990
33	H.R. 918: Bill Would Stop Mining on Our Public Lands	1992
24	History of U.S. Mining Law, The	1990
23	Is "Activism" a Dirty Word for Miners?	1992
17	Mineral Policy Center Launches Another Attack on Miners	1993
15	Mineral Policy Center Supports Environmental Extremism	1990
9	Miners Are a Threatened and Endangered Species	1989
21	Miners Dominate Reno Mining Law Hearing	1991
20	Miners Show Strength and Unity at Mining Law Hearing	1990
11	Mining Issues in 1993: What Can We Expect?	1993
2	Mining vs. Environmentalism	1985
18	Mother Nature Charged with Environmental Crimes	1993
16	MPC Leads Effort to Abolish U.S. Mining Laws	1991
3	Next Mining Boom, The	1985

Continued

Article No.	Article Title in Alphabetical Order	Pub. Year
4	Placer Mining vs. Natural Erosion	1985
35	Recent Action on Mining Laws in Congress	1992
32	Rep. les AuCoin Launches Assault on Miners	1992
27	Rep. Rahall Introduces New Mining Law Bill	1991
36	Senator Bumpers Launches S.257 Anti-mining Missile	1993
25	Senator Bumpers' S. 1126 and Potential Impacts of Similar Legislation	1990
38	Senator Craig's S. 775 Clears Senate Committee	1993
22	Strong Opposition Shown Towards Sen. Bumpers' Sen.433	1991
5	Wilderness Issue, The	1985
10	Wilderness Update 1991: How Much Is Enough?	1991

APPENDIX C

Author's Comments on Bumpers' Bill S.1126

Editor's Note: The following content was extracted from *article 25*, "Sen. Bumpers' S. 1126 and the Potential Effects of Similar Legislation," on page 131 of this volume. It contains a brief description of each of the sections within the titles in the bill.

Title I—Definitions

Sections 102 (1) and (2): The definitions under "contiguous group of claims" and "contiguous group of patents" requires a single operator and operation. It does not allow for either exploration or development by joint ventures, partnerships of individuals operating on an equal basis, nor does it allow for multiple mineral developments within the same operating area unless it is under a single operator. Does not allow different types of mining operations in the same area, and generally does not address the realities involved in either hardrock mineral exploration or development.

Section 102(7): The definition of "locatable lands" withdraws certain types of lands that are presently open to mineral entry.

Title II—Disposition of Mineral Deposits

Section 201 Prospecting: All prospectors would be required to notify the federal management agencies before initiating mineral exploration. This would result in loss of confidentiality and could result in a requirement to report on prospecting activities and any results obtained prior to the filing of an exploration claim. It might also allow the agency to refuse to let prospectors enter certain areas, or restrict their activities.

Section 202 Exploration Claims (a): Allows for only 20-acre exploration claims that conform to public land survey system wherever possible. Eliminates lode and placer distinction as well as larger-sized claims currently allowed for placers. It does not provide for fractional claims, and it leaves open the possibility that new filing and locating requirements can be formulated by federal agencies.

Section 202(b): States that new federal location and filing requirements take precedence over state and county requirements, and stipulates a $100 recording fee per claim. Also authorizes formulation of new regulations that would replace the current 43 CFR 3809 regulations, and eliminates all extralateral mining rights. Claim can be used only for exploration.

Section 202(d)(1): Requires submission of an "exploration plan" to federal agencies prior to making any significant surface disturbance. Gives the managing agency the right to disapprove the plan, at which time the new claim would become null and void by operation of law. Does not allow for an appeal process, and gives federal agency total discretion.

Section 202(d)(2): Requires submittal of a reclamation plan along with the exploration plan, which makes it likely that even minimal exploration activities will have to be bonded. Locator can only take samples for analysis, cannot remove minerals for sale, and can only conduct activities required for exploration for hardrock minerals. By exclusion, this section also eliminates exploration for placer deposits and disallows activities such as dredging and sluicing on exploration claims.

Section 202(e): Again disallows any activity other than exploration, and imposes a fine of not more than $1,000 per day for any violations and voids the exploration claim at the same time. In other words, the claims cannot be used for recreational or camping purposes—only for mineral exploration for hardrock minerals.

Section 203 Mineral Patents (a): Prior to any mineral development or production, the locator of an exploration claim must have an approved exploration plan (and reclamation plan) and submit an application for a mineral patent. This means that a mineral deposit must be discovered and be proven to be economically feasible to mine before any mining can take place.

Section 203(b): States that a claim is presumed to be abandoned and null and void after 10 years unless the locator has applied for a mineral patent. The patenting process could take years to complete.

Section 203(c)(1) and (2): Federal agency has to determine that a hardrock mineral deposit capable of commercial development (economically feasible) exists on the property and that an approved mining and reclamation plan are in place before a mineral patent can be issued. This would imply that marginal hardrock, all placer and locatable industrial mineral deposits cannot be patented or mined under any conditions, and it leaves the question of patent issuance to the discretion of the federal government.

Section 203(d): Only allows for exclusive patenting of hardrock minerals and, by exclusion, eliminates all placer and industrial mineral development. Also does not allow for leasable mineral development (or exploration). It stipulates a surface use fee of not less than $5 per acre and an overriding federal production royalty of 8% of the value of hardrock minerals. States patent is void if mineral production has not begun within 15 years. In other words, a mining patent would only be a long-term lease.

Section 203(e): Stipulates that lands patented for minerals cannot be used for any purposes other than mining. Violations would be subject to a fine of not more than $1,000 per day and the mineral patent would be voided.

Section 203(f)(1-11): States the requirements for a mining and reclamation plan (allowing for additional limitations), including information that is unnecessary, unavailable or meaningless. By structure, this section allows for overly stringent and unrealistic regulation by which a mining operation could be shut down at any time, at the discretion of the government.

Section 204 Diligence (a)(1-3): Requires a minimal expenditure each year of $50 per acre for each exploration claim and $100 per acre for each mineral patent during years 1 through 5; requires expenditures of $100 per acre for exploration claims and $200 per acre for mineral patents during years 5 through 10; and requires expenditures of $300 per acre on mineral patents during years 10 through 15. This means that diligence (assessment) costs per claim per year would amount to: (1) $1,000 per exploration claim and $2,000 per mineral patent in years 1-5; (2) $2,000 per exploration claim and $4,000 per mineral patent in years 5-10, and (3) $6,000 per mineral patent in years 10-15. These amounts would put the small hardrock miner out of business almost immediately, would severely impact exploration companies, and (for existing claims) would put dredgers and small placer miners out of business within a short period of time.

Section 204(b): Allows cash payments of amounts listed in Sec. 204(a)(1-3) above in lieu of performing annual labor. Considering the high amount of such fees, this provision is meaningless for small miners and exploration companies—they would be priced out of business. In addition, in-lieu payments do not encourage the exploration and discovery of minerals.

Section 204(d): Annual payments in lieu of labor expenditures are not required after production is reached and the 8% overriding federal royalty is being paid. This also is meaningless, because an 8% royalty on production is roughly equivalent to at least a 16% (or more) tax on net profits, which would be in addition to existing state and federal taxes.

Section 204(e): Requires payment of an additional $100 "holding fee" each and every year for each exploration claim or mineral patent, as well as filing proof of annual labor of payment in lieu of labor. The agency is required to determine that the work has been performed (inspections) or the fees paid every year and, if these requirements are not met, the claim or mineral patent will be declared null and void by operation of law.

Section 205 Use of Federal Surface: Allows claimant to submit application for surface use on lands other than exploration claims or mineral patents in support of mineral development and production. An annual surface use fee of $5 per acre is again required, as well as reclamation plans (most likely bonded) and other terms and conditions as yet unspecified.

Section 206 Federal Reservation of Rights: Reserves all rights and interest in locatable lands to the federal government except for hardrock minerals covered by a valid mineral patent, subject to the overriding 8% royalty. This means that only the right to explore for minerals is obtained by filing an exploration claim—no rights to minerals are obtained until patented.

Title III—Environmental Protection

Section 301 Regulations and Reclamation (a)(b): Requires federal agencies to issue new regulations covering all prospecting, exploration, development and production that will address all potential environmental concerns. It stipulates that all mined areas will be reclaimed and restored to a condition supporting the original use prior to mining, with concurrent reclamation when feasible. This section lists standards to be addressed, including topsoil replacement and open pit backfilling—which would practically eliminate open mining operations unless exceptions were granted.

Section 302 Bonds: Requires the owner of an exploration claim or mineral patent to post a performance bond sufficient to ensure complete and timely reclamation. Since the bond is conditioned upon compliance with this Act and all other laws and regulations, no bonding company or financial institution would consider issuing such a bond. As a result, cash bonds or collateral would be required—eliminating small mines, marginal operations and exploration companies. If no bond can be obtained, then no mining activity can take place.

Section 303 Amendments to Land and Resource Management Plans: Provides that federal agencies will address mining and exploration activities in the preparation, revision or amendment of land and resource management plans, providing for additional terms and conditions and allowing prohibition of certain types and classes of activities. This is a wide-open invitation to inflict stringent and overly restrictive regulations on mining activities such that they would effectively prohibit mining from taking place.

Title IV—Disposition of Receipts

Section 401 General (a) and (b): Provides that federal government and states in which mining is located shall split surface use fees, payments in lieu, and overriding royalties 50/50—which is a major portion of the bill. It allows onerous industry-specific taxation of mining with the express purpose of obtaining money that will be used for other purposes and will not benefit mining in any way. Also provides that recordation fees and annual holding fees will go to the managing agency for the purpose of administering this Act—read "enforcement." In other words, miners would be paying the federal government to put themselves out of business.

Title V—Existing Claims

Section 501 New Claims and Patents (a): Repeals the Mining Law of 1872 and all amendments thereto for new mining claims and requires that they be located under this Act.

Section 501(b): Eliminates the right to patent mining claims located under the 1872 Mining Law. Effectively takes away valid existing rights.

Section 502 Existing Claims (a): Requires owners of valid existing claims to make one of two choices within 3 years on enactment of this Act. The failure to do so automatically invalidates their claims.

Section 502(b)(1-2): First Choice—Owner may relocate claim under the requirements of this Act, and relocation must be completed within 60 days after making the choice. The claim would then be administered pursuant to the Act except: the annual holding fee and annual surface use fee would not apply to the relocated claim, and the annual labor or payments in lieu would be one-half of the amounts specified in Section 204. This option would take away the mineral rights previously held by the owner under the 1872 Mining Law as well as any grandfathered rights (surface use, etc.). The new claim would be an exploration claim under the Act.

Section 502(c)(1-3): Second Choice—Owner may choose to maintain claims by complying with mining laws in effect on the date this Act is enacted. However, 3 years after the date of enactment of this provision the assessment work or payment in lieu of labor requirement would be fixed at $5,000 per mining claim per year. In addition, full reclamation and the posting of a bond for same would be required according to Sections 301 and 302 of the Act.

APPENDIX D

Description and Analysis of Rahall's Bill HR 3866

Editor's Note: The following text, consisting of descriptions of the sections in Rep. Rahall's bill H.R. 3866, was extracted from **article 26,** "HR 3866: Rahall's 'Mineral Exploration and Development Act of 1990." The article begins on page 135.

Section 1: States the short title of the legislation.

Section 2: Contains definitions and references used in the bill, most of which are self-explanatory. However, one notable addition is "diligence year," which means the 365-day period commencing on the date a mining claim is recorded with the Secretary of the Interior and each 365-day interval thereafter. The diligence year would replace the assessment year and annual calendar year filings currently in force, so each claim would be subject to filing requirements dating from the time it is recorded.

Section 3: States that a mining claimholder's rights include an exclusive right of possession for mineral exploration, development and all directly related activities, but that no other uses are permitted.

Section 4: Describes procedures for locating and recording mining claims on survey lands, protracted survey lands, and unsurveyed lands. Notice of location posted on northeast corner of claim; claims will be 40 acres in size and conform to legal subdivisions of survey system where possible; 40-acre claims on unsurveyed lands will be regular square with monuments at each corner; description on notice of location will determine boundaries of claims; claims must be recorded within 30 days after location; serial number and certificate to be issued by agency upon recording (maintenance requirements on certificate); claim only covers mineral deposits within the boundaries of the claim (no extralateral rights); rival claimant can locate claim over existing claim if it can be proven that requirements have not been met; and claimant who fails to comply with requirements or drops mining claims cannot file a new claim on that site for six months.

Section 5: Outlines claim maintenance requirements. Claimant must pay the rental fees and either meet diligent expenditure requirement or payment in lieu of such expenditure. Claim is forfeited if requirements not met; the diligent expenditure is not required when an approved mining plan has been issued or when claimant is prevented from doing work by legal actions.

The required rental fee will be a minimum of $1.50 per acre per year until a mining plan is approved, at which time it will increase to not less than $5 per acre per year. Fees must be paid at the time the affidavit on diligent expenditures is filed each "diligence year" (annually). Rental fees will go into a new fund for reclaiming past mining disturbances.

Diligent expenditures may be made on one claim in a contiguous group, and include, but are not limited to, activities such as: investigations and surveys (geotechnical, geological, geophysical and geochemical); bulk mineral sampling and testing; drilling, environmental and engineering studies, and reclamation or restoration activities. Minimum annual expenditures shall be $20 per acre in each of the first through fifth diligence years; $40 per acre in each of the sixth through tenth years; $80 per acre in each of the eleventh through fifteenth years; and $160 per acre in each of the sixteenth through twentieth years and every year thereafter.

An annual affidavit must be filed that contains clear and convincing proof of the value and beneficial nature of diligent expenditures, with sufficient detail to provide for validation of same. After the fifth year, a claimant may elect to make cash payments in lieu of expenditures, as follows: $20 per acre in years 6-10; $40 per acre in 11-15; and $80 per acre in years 16-20 and thereafter. Documentary proof of diligent expenditures reported on affidavits must be maintained for a period of 5 years.

A claimant with an approved mining plan must submit a report at least annually demonstrating bona fide efforts to produce minerals. If it is determined that these efforts are not sufficient, the claimant must comply with the diligent expenditure requirement. Outlines conditions when mineral production is suspended.

Claimholder is required to file a document containing his name and address and serial numbers of each claim on or before the anniversary date of each diligence year, accompanied by the required rental fees and the affidavit for diligent expenditures. Failure to file these instruments will result in voiding of claims. The federal agency may audit claimholders as necessary to ensure compliance with requirements.

Civil penalties of not more than $500 per violation for each day of such violation will be assessed if a claimant submits false, inaccurate or misleading information on diligent development expenditures on affidavits or fails to file bona fide efforts report (when required). Penalties do not apply if violation is corrected within 20 days after receiving notification of such violation or if reported and corrected by claimant within 20 days. Claimants who: (1) knowingly or willingly prepare, maintain or submit false, inaccurate or misleading information on diligent expenditures on affidavits, or, (2) knowingly or willfully fail to file the bona fide efforts report (if required), or (3) fail or refuse to permit an audit of claim records, are liable to a penalty of not more than $1,000 per violation per day of such violation if it continues after being given due notice by the government. Civil penalties will not be assessed until the claimant charged with a violation has been given an opportunity for a hearing, and findings may be appealed in U.S. District Courts. Other conditions related to civil penalties are also outlined, including judicial review.

Section 6: Requires the Department of the Interior to promulgate rules, regulations and forms, including:

- Standardized form for Notice of Location.
- Standardized form of recorded certificate.
- Standards and guidelines defining diligent expenditures that qualify.
- Standardized form for affidavits filed on diligent expenditures.
- Formulation of procedures to verify expenditures.
- Standardized form for filing of annual instruments.
- Standardized form for filing bona fide efforts report.
- Rule guidelines to determine bona fide efforts.
- Rule guidelines concerning evidence required to establish failure to comply with diligent expenditure.
- Establishment and collection of user fees to cover the cost of administering the Act's requirements.

Section also states that this Act will preempt any conflicting state or local requirements, but allows states to enact reclamation and environmental laws that are stricter than federal requirements.

Section 7: States that, after Jan. 23, 1990, no patent shall be issued for mining claims or mill sites located under the General Mining Laws unless a patent application was filed prior to that date and all requirements for the patent application have been met.

Section 8: Requires that all mineral activity be conducted in such a way as to prevent unnecessary degradation and minimize adverse environmental impacts. Requires claimants to file plans of operations for any activities that may result in a significant surface disturbance and file mining plans for any proposed mining, milling, processing, and related activities. It requires financial or other guarantees for all types of plans to ensure reclamation (mandatory bonding). No plans may be approved for claimants who have failed to comply with surface management requirements or standards.

The Department of the Interior is required to establish and enforce reclamation standards covering all relevant concerns and conditions. Also authorizes at least quarterly (every three months) inspections of mining claims to ensure compliance with plans, and allows any person who is or may be affected by mineral activities to notify the government of any violations which he or she has reason to believe may exist at a mining site. The Department may utilize enforcement personnel from the Office of Surface Mining Reclamation and Enforcement Agency to augment BLM and USFS personnel for inspections and enforcement.

Claimants who fail to comply with surface management regulations, any statutes, rules or regulations, and other new standards will be liable to a civil penalty of not more than $1,000 per violation for each day of such violation after the date a notice is issued by the government. States that all existing regulations not in conflict with the Act will still apply.

Section 9: States that certain provisions of previous Acts will also apply to multiple mineral development, surface resources and enforcement.

Section 10: Eliminates industrial mineral materials from location of mining claims after H.R. 3866 is enacted, including uncommon varieties currently locatable under the Mining Law. States that industrial minerals will only be salable and leasable in the future. Also clarifies disposal of mineral materials by amending various sections and language in 30 U.S.C. 612 and 30 U.S.C. 601 (and following).

Section 11: Requires that all land use plans prepared by land managing agencies will include numerous specific requirements, with the primary purpose of protecting non-mineral values. Also requires that existing land use plans be amended to comply with the Act and subsequent new regulations.

Section 12: Transfers the authority to administer mining activities on National Forest Service lands from the Bureau of Land Management (Department of the Interior) to the Department of Agriculture. Outlines requirements and procedures to accomplish this transfer in authority.

Section 13: Directs the Department of the Interior to formulate and issue the final regulations to implement this Act within 180 days after the Act is enacted, with the regulations being effective upon publication in the Federal Register. Notice of the requirements of this Act must be sent to existing mining claimants within 30 days after publication.

Section 14: States that all mining claims must be located according to the provisions of this act after 180 days following enactment of same. Before the date 3 years following this date, claimants may elect to convert their claims by filing a new notice of location as specified in the Act, at which time the claims would fall under the provisions of the Act.

States that, on the date 3 years after the date 180 days following the enactment of this bill, all unconverted mining claims located under the 1872 Mining Law will be null and void. In other words, all existing mining claims must be converted or they will become invalid.

States that lode, placer, tunnel site, etc. distinctions are eliminated on claims located under the Act, and that "discovery" is eliminated as a requirement once a claim is subject to the Act.

Also states that claims may not be recorded under this Act during the 3-year period allowed for conversion if they cover lands upon which prior existing claims are being maintained under the General Mining Laws.

References

AuCoin, L. (1992). Sodium cyanide mining: the drift nets of our public lands. Dear Colleague letter introducing H.R. 4298.

Bumpers, D. (Autumn 1988). Mining law's flaws—a congressional assessment. *Clementine: The Journal of Responsible Mineral Development*. Reprinted from the Congressional Record, October 18, 1988.

Fitzgerald, R. (April 21, 1991). The great train robbery. *Reader's Digest*

Gardiner, D. (1992). Sierra Club news release.

Hocker, P. (Autumn 1988). An introduction to the Mineral Policy Center. *Clementine: The Journal of Responsible Minerals Development*.

Hocker, P. (Autumn 1989). Cyanide spring, heaps of gold, pools of poison. *Clementine: The Journal of Responsible Minerals Development*.

Holmes, D. (November 1991). *Northwest Mining Association Bulletin*.

Mines of poison. (Autumn 1988). *Clementine: The Journal of Responsible Mineral Development*.

New gold rush, the (n.d.). ABC News *20/20* videotape.

Our land. (1988). *Blueprint for the Environment—Advice to the President-elect (George Bush) from America's Environmental Community*.

Satchell, M. (October 28, 1991). The new gold rush. *U.S. News & World Report*.

Strobel, P., ed. (November/December 1991). *Mining World News*.

Accuracy in Media. (Sept.-A 1989) *20/20 Turns Toxic* report.

Turque, B. (September 30, 1991). The war on the west. *Newsweek*.

Udall, S. (Autumn 1988). *Clementine: The Journal of Responsible Mineral Development*.

Index

$100 mining claim holding fee, 49, 53, 57, 157, 179, 180, 181, 186
8% gross production royalty, 89, 170, 184, 185-186, 189, 190, 196, 206, 207
12.5% gross production royalty, 94, 156, 196
1872 Mining Law
  defense of, 111-113, 115-116, 117, 119-120, 168
  intent of and changes to, 26, 127-129
  legislation regarding, H.R. 322, Rahall: 185, 189-193, 195, 196; H.R. 918, Rahall: 115-118, 141-142, 147-150, 158, 167-171, 185, 194, 196; H.R. 918 "substitute," Rahall: 50, 167-171, 173-177, 185, 189, 196; H.R. 1096, 151-154, 159; H.R. 2614, DeFazio: 155-159; H.R. 3866, Rahall: vii, 111, 112, 135-138, 141, 171, 194, 211-214; H.R. 4298, AuCoin: 161-165; S. 257, Bumpers: 183-187, 195, 196; S. 433, Bumpers: 119-120, 143-146, 155, 157, 158, 159, 181; S. 433 "amended," Bumpers: 173-176, 186; S. 775, Craig: 195, 197; S. 785, Burns: 119; S. 1126, Bumpers: 81, 111, 131-133, 143, 205-209
  opposition to and efforts to change, 30, 41, 49, 50-51, 52, 53, 73-79, 81-85, 87-89, 91-94, 108, 115-118, 131-133, 135-139, 143-146, 150, 155-159, 170, 173-177, 179-181, 183-187, 189-193, 195-197
1987, significant mining-related events and trends during, 33-35
abandoned mines, 91-94
Abandoned Mine Lands program, Nevada, 95
ACECs, 57

Alaska, wilderness lands in, 46, 71, 98, 101, 153, 183
Amendment in the Nature of a Substitute for H.R. 918, 167-171, 189
  See also Rahall's bill H.R. 918 "substitute" bill
AML, Nevada, 95
anti-mining activists, projects of, 73
anti-mining bias in media, 45, 121, 180
anti-mining misrepresentations and misinformation, countering, 87, 89
anti-mining propaganda, 105-110
Areas of Critical Environmental Concern, 57, 152, 153
AuCoin's H.R. 4298, 161-165

Babbitt, Secretary of the Interior Bruce, 91, 94, 196
Bingaman, Sen. Jeff on Bumpers' bill S. 433, 120
bonding. *See* reclamation and bonding
Bryan, Sen. Richard, 119
Bumpers/Miller/Rahall plan of 1992, 196
Bumpers, Sen. Dale
  on 1872 Mining Law, 75
  on abandoned mines, 91
  on Sen. Craig's S. 775 bill, 195, 196
  vs. Sen. Harry Reid's amendment, 179, 180
Bumpers' bill S. 257, 183-187, 195, 196
Bumpers' bill S. 433, 119-120, 143-146, 155, 157, 159, 181. *See also* DeFazio's bill H.R. 2614
Bumpers' bill S. 433, "amended," 173-176
Bumpers' bill S. 433, hearing on, 119-120
Bumpers' bill S. 1126, 81, 111, 131-133, 136, 143
  analysis of, 205-209

Bumpers' Trojan horse bill, 173-177, 186, 196. *See also* Bumpers' bill S. 257
Burden of Gilt report, 91-95
Bureau of Land Management
   mining regulations, 57
   reclamation bonding. *See* reclamation and bonding
bureaucratic terminology, satire on, 61-63
Burns' bill S. 785, 119

citizen suits against mining, 145, 159, 167-168, 174, 193
civil litigation as anti-development and anti-mining tactic, 29, 43, 85
civil penalties, proposal for in mining law "reform," 56, 136, 145, 158, 159, 174, 212, 213
claim holding fee, annual mining, 49, 53, 57, 157, 174, 186
Clementine journal, 74, 75, 81, 84, 85, 87-88
Clinton Administration on mining law "reform" strategy, 195, 196
Clinton, President on Bumpers' S. 257, 195-196
conservation movement, origins of and co-opting of by radicals, 69, 70
Craig, Rep. Larry on 1872 Mining Law, 112
Craig's bill S. 775, 195-196
criminal penalties, 56, 145, 158, 159, 174, 212, 213
cyanide
   and environmentalists' fear-mongering, 84, 121, 161-165
   as surrogate issue, 165
Cyanide Mining Impact Relief Act of 1992, 161-164

Darden, Rep. George "Buddy," 152
DeConcini, Sen. Dennis on Bumpers' bill S. 433, 120
DeFazio's H.R. 2614, 155-159. *See also* Bumpers' S. 433

DeFazio's amendment to H.R. 918 "substitute" bill, 169
Defenders of Wildlife, 161, 163
Dougherty, James, 163

Endangered Species Act (ESA), 51, 56, 99
environmental activities, geological events vs. human, 97-104
environmental concerns
   and influence of extremists regarding, 21-24, 77-78, 143
   federal legislation to address, 128, 151, 152-153, 168
environmental extremism and radicalism, 11-14, 21, 25, 26, 27, 33, 34, 38, 41-44, 49, 50-54, 57, 58-59, 67-72, 87-89, 92, 97-103, 106, 152, 153, 155, 159, 161, 162, 167-171, 175-176, 185-187, 189, 192, 195
environmental extremism, dangers inherent in, 43
environmental extremist activists
   appeal to emotion, 42
   individuals and organizations, 74-76
   media bias and propaganda, 45, 87, 88, 89, 121, 180
   objectives, strategies and tactics of, 29-30, 41, 49, 50, 55, 58, 67, 73, 74-77, 78, 87, 88, 89, 106, 192
   special treatment of politically, 87, 88-89
Environmental Impact Statements, 44, 145, 192
Environmental Impact Studies as anti-mining tactic, 29, 43
environmental laws and regulations, stringent state, 57-58
environmental legislation, extremist, 51-52, 54
environmental movement
   leaders of, 74
   purposes and tactics of, 77-79
   radicalization of, 73

Index

environmental protection mania, 14, 31, 43, 44, 52, 54, 72, 85
environmental protectionism, 12, 29-31
EPA regulations and overregulation, 21, 22, 54-55, 58. *See also* U.S. mining, overregulation of

Federal Environmental Protection Agency, 54, 97
Federal Land Policy and Management Act, 128, 151-153
Fields, Russ, 116, 118. *See also* Rahall's bill H.R. 918, field hearing on
final rulemaking by government agencies, 76
fish and wildlife, effects of placer mining on, 21, 23-24
FLPMA. *See* Federal Land Policy and Management Act

Gardiner, David, 165
General Mining Law of 1872, 128. *See also* 1872 Mining Law
gold miners, legislative and media assaults on, 161-165
gold rushes in western U.S., role in U.S. Mining Law enactment, 127
government policies and regulations vs. mining, 34, 38, 41
government regulations, excessive and unrealistic, 76. *See also* U.S. mining, overregulation of
gross production royalty, 94, 144, 156, 184, 185-186, 189-190, 196

H.R. 322. *See* Rahall's bill H.R. 322
H.R. 918. *See* Rahall's bill H.R. 918
H.R. 918 "substitute" bill. *See* Rahall's bill H.R. 918 "substitute"
H.R. 1096. *See* Vento's bill H.R. 1096
H.R. 2614. *See* DeFazio's bill H.R. 2614
H.R. 3866. *See* Rahall's bill H.R. 3866
H.R. 3866, hearing on. *See* Rahall's bill H.R. 3866, hearing on
HAMR, 92, 93, 94

Hardrock Abandoned Mines Reclamation Program, 92, 93, 94
hardrock minerals, royalty on, 87, 119, 120, 156, 184, 185-186, 189-190, 196, 206, 207
Hocker, Phil, 74-75, 81, 84, 88, 92, 93, 94, 95, 111, 163, 164, 167, 196
Hocker, Phil M. *See* Hocker, Phil
Hocker, Philip M. *See* Hocker, Phil
Hohmann, Kathryn, 195
Holmberg, Patricia A., 117

Idaho Mining Association, 117
Ingle, Jr., Hugh C., 117, 118
Interior appropriations bill FY1993, mining law proposals in, 179-180
Interior, Department of, 26, 46, 91, 108, 149, 169, 179, 212, 213, 214

Jamison, BLM Director Cy, 112, 120, 151, 152
Johnston, Sen. J. Bennett, 195

land management policies and public lands/natural resources, 47-48
land restrictions and withdrawals, 57
land use planning and review processes, use of against miners, 191
lawsuits, abuse of by anti-mining factions, 58
Livermore, John, 117. *See also* Rahall's bill H.R. 918, field hearing on
localized viewpoint, 22
Lode Law of 1866, 127-128
Lujan, Jr., Manuel, Secretary of the Interior, 168

Madigan, Department of Agriculture Secretary Edward, 168
Marlenee, Rep. Ron, 152
Miller, Glenn, C., of Sierra Club, 109
Miller, Nevada Governor Bob, vii, 112, 115-116, 169
Miller, Rep. George, 51, 91-92, 94, 142, 167, 168, 175, 181, 189, 196

219

Mine Free by '93 slogan, 50, 81, 175, 185, 187. *See also* Sierra Club
Mine Remediation Trust Fund, 164
Mineral Exploration and Development Act of 1990, vii, 111, 135, 211. *See also* Rahall's bill H.R. 3866
Mineral Exploration and Development Act of 1991, 115, 141, 147, 150, 161. *See also* Rahall's bill H.R. 918
Mineral Exploration and Development Act of 1993. *See also* Rahall's bill S. 257
mineral exploration and development in U.S., government policies and regulations harmful to, 50, 53
Mineral Leasing Act of 1920, 75, 128
Mineral Policy Center and environmental extremism, 74-75, 76, 81-85, 87-90, 91-95, 111, 121, 161, 163-164, 167, 196
minerals industry in America, need for strong, 7-10
minerals supply in U.S., environmentalists' influence on, 11-14
minerals supply, obtaining domestically vs. import reliance, 18
miners and mining, negative perceptions of, 41, 42-43, 49, 55-56
mining activism, need for. *See* pro-mining activism
mining activity, effects of mining industry's perceptions on, 60
mining business, impacts of negative propaganda on, 107-108
mining claim holding fee, 49, 53, 57, 157, 179, 180, 181, 186
mining claim patenting, 52-53, 88, 109, 112, 128, 135, 137, 149, 155-156, 179-181, 205, 206, 207, 208
mining industry, attacks on and threats to, 41-44, 49, 55-56, 67-72, 81-85, 91-94, 105-110, 161-164, 167-171, 183-187, 189-193, 195
mining industry, boom and bust aspects of in early 1980s, 15-19

mining industry's need for activism and positive publicity, 42, 73, 79
mining issues in 1993, 49-60
Mining Law Act of 1980. *See* Bumpers' bill S. 1126
mining law hearing on Rahall's bill H.R. 3866, 111-113, 147
Mining Law of 1872. *See* 1872 Mining Law
mining law proposals, U.S. Senate actions August 5, 1992, 179-180
Mining Law Reform Act of 1991
  House bill, 155. *See also* DeFazio's bill H.R. 2614
  Senate bill, 119, 143. *See also* Bumpers' bill S. 433
mining laws, targeting of federal, 41
mining vs. radical environmentalism, 11-14
Mining World News, 106-107, 121
mining, lawsuits against as a tactic of anti-mining entities, 58
mining, media bias against, 105-109
mining, public attacks on, 59
mining, regulation of in U.S. *See* U.S. mining, overregulation of
Mother Nature
  and survival of the fittest, 56, 99
  and the environment, 97-104
MPC. *See* Mineral Policy Center
multiple use concept vs. environmental extremism, 12, 27, 29-30, 41-42, 45, 47-48, 71-72, 151

Nader, Ralph, 111
National Conservation Areas, 57
National Environmental Policy Act, 128, 191
National Recreation Areas, 27, 47, 57
National Wildlife Federation, 74
natural resource development, hostility and threats to, 8, 11, 13, 43, 45, 50, 54, 55, 57, 58, 59, 69, 70, 72, 73-79, 151, 152-153, 154
natural resources management, 12, 27

Index

natural resources, impact of withdrawals on availability of, 45-48
NCAs (National Conservation Areas), 57
NEPA process, 192
Nevada Abandoned Mine Lands program, 95
Nevada Miners and Prospectors Association, vii, 117
NRAs (National Recreation Areas), 57
news media
 anti-mining propaganda, 52-53, 59, 87, 88, 105-107
 attacks on mining, counteracting, 106-109
 bias in favor of environmentalism, 13, 105-107, 180
 manipulation of by environmental activists, 42
 reports on mining, fact vs. propaganda, 108-109, 197-198
 targeting of mining by, 41

overregulation as national security threat, 31

Parkhurst, Dave W., ix, 3-4, 117, 199, 200
patenting of public lands. *See* mining claim patenting
patents, mineral/mining. *See* mining claim patenting
People for the West!, 115. *See also* Rahall's bill H.R. 918 field hearing
perpetual litigation as anti-development strategy, 30, 143, 145, 159
Pickett Act of 1910, 128
Placer Act enactment in 1870, 128
placer disturbances, treatment of as unnatural, 23
placer mining vs. natural erosion, 21-24
placer operations, high visibility of, 21-22
platinum group metals, 8, 17
politicians
 anti-mining actions by, 49, 50
 targeting of mining by, 41

pot-holing effect, 23
Potash Leasing Act of 1917, 128
pro-mining activism, 42-43, 44, 121-123
public domain lands in U.S.
 in Western vs. non-Western states, 183
 multiple use vs. environmental elitism and extremism, 45, 47-48
 Rahall's H.R. 918 "substitute" bill, projected impacts of, 167-168, 189
 wilderness from and restrictions on, 27, 71, 72, 73, 187
public land multiple use vs. wilderness withdrawals and land restrictions, 29, 41, 71

radical environmentalism. *See* environmental extremism and radicalism
Rahall, Rep. Nick on mining law "reform," 91-92
Rahall's bill H.R. 322, 185, 189-193, 195, 196
Rahall's bill H.R. 918, 88, 115-118, 141-142, 147-150, 158, 167, 181, 185, 194, 196
Rahall's bill H.R. 918, field hearing on, 115-118
Rahall's bill H.R. 918 "substitute" bill, 50, 167-171, 173, 174, 175, 176, 181, 185, 189, 196. *See also* Amendment in the Nature of a Substitute for H.R. 918
Rahall's bill H.R. 3866, 111, 112, 135-139, 141, 147, 150, 171, 194
 analysis of, 211-214
rare earth metals, 17
RCRA, 51
reclamation and bonding, 37-40, 82, 92-93, 163, 192-193
Red, White 'n Blue Mining Act, 81-83
regulatory excess. *See* U.S. mining, overregulation of
Reid, Harry amendment, mineral patent provisions of, 179

Resource Conservation and Recovery
Act, 51, 128

S. 257. *See* Bumpers' bill S. 257
S. 433. *See* Bumpers' bill S. 433
S. 1126. *See* Bumpers' bill S. 1126
S. 775. *See* Craig's bill S. 775
S. 785. *See* Burns' bill S. 785
Satchell, Michael, 109
sensitive lands, designation of as
    protected wilderness, 45
settlement of American West, role of
    U.S. Mining Law in, 127, 128
Sierra Club
  activism of and leaders in, 74-75,
    109, 121, 161, 164, 195
  slogan "Mine Free by '93," 50, 81,
    175, 185, 187, 189
Simpson, Sen. Alan on Bumpers' bill
    S. 433, 120
small mining and prospecting, threats
    to, 132
Smith, Don, 117. *See also* Rahall's bill
    H.R. 918, field hearing on
Spanish Royal Code of 1783, basis for
    American West mining law, 127
Stevens, Ted amendment on mining
    claim holding fee, 180
Strobel, Paul S. of *Mining World News*,
    106-107
Surface Resource Act of 1955, 128

Task Force on Federally Owned Mineral
    Lands in 1976, 26
taxation, selective, 94
Taylor, Mack, 117. *See also* Rahall's bill
    on H.R. 918, field hearing on
Title 30 of United States Code, 129
Title II of Rahall's bill H.R. 322, onerous
    provisions of, 189-193
Title II of Rahall's bill H.R. 918, 181
Trojan horse bill, 12, 186. *See also*
    Bumpers' bill S. 433

U.S. economy, threats of anti-mining
    legislation to, 133, 146
U.S. Forest Service lands, transfer of
    responsibility for, 149
U.S. Mining Law
  effect of environmental concerns on
    changes to, 128
  history and development of, 127-129
U.S. mining laws, anti-mining activists
    organizing against, 73-79
U.S. mining, overregulation of, 21, 30-
    31, 33, 34, 37, 41, 43, 49, 50, 76,
    97-103, 144, 158. *See also* EPA
    regulations and overregulation
Udall, Stewart L., 74, 81. *See also*
    environmental movement
Ugalde, Sammye, 116. *See also* Rahall's
    bill H.R. 918, field hearing on

Vento's H.R. 1096, 151-154
Vento, Rep. Bruce, co-sponsor of
    Rahall's bills H.R. 322 and H.R. 918,
    142, 151, 189
Vucanovich, Rep. Barbara, 112, 117,
    118, 119-120

western states, effects of Vento's bill
    H.R. 1096 on, 151, 154
wilderness buffer zones, 30, 71, 73, 77
wilderness issue in U.S., 25-27, 45-48
wilderness legislation in Congress, 54
wilderness system in U.S., role of gov-
    ernment agencies in managing, 27
wilderness withdrawals and land
    restrictions
  adverse effects of, 11, 12, 13, 71-72
  massive designations of public lands
    as, 26, 27, 45, 46, 54, 57, 73, 74
  vs. multiple use concept, 45, 47-48

Wirth, Sen. Tim on Bumpers' bill S. 433,
    120
world economy's impact on mining, 59

Young, Rep. Don on Vento's H.R. 1096,
    152, 154, 159

Editor Contact Information

Susan Lee (Sue) Parkhurst is the publisher/editor of Pine Nut Press in Minden, Nevada. To learn more about the publications and services offered by Pine Nut Press, or to see more of Dave's writing or his project photos, please visit **www.pinenutpress.com**.

The print version of the Dave W. Parkhurst Mining Writing Collection may be purchased from Amazon.com. E-book versions of the volumes are planned and will also be available at Amazon.com.

www.ingramcontent.com/pod-product-compliance
Lightning Source LLC
Chambersburg PA
CBHW020639220526
45464CB00001B/220